From Demo to Delivery:
The Process of Production

Contents

Section A • Preparation

Section B • The Sessions

v

Contributors

Paul Allen

Paul Allen, associate professor at Middle Tennessee State University, earned his B.S. in business management from the University of Tennessee. He was awarded the M.B.A. with an emphasis in marketing from MTSU. Additional studies have included courses and seminars at Vanderbilt, Harvard, Clemson, and Belmont Universities. Paul began teaching at MTSU in 1999 as an adjunct faculty member in marketing of recordings for the recording industry program. Paul Allen's career has included work in the radio and television industries, as well as being executive director of Country Radio Broadcasters, Inc., an industry trade association. He has been producer or executive producer for scores of stage productions for acts including Alan Jackson, Garth Brooks, the Dixie Chicks, Toby Keith, Trace Adkins, Trisha Yearwood, Clint Black, Vince Gill, and Martina McBride in conjunction with the annual Country Radio Seminar. For seven years he was the executive producer of the New Faces of Country Music Show presented in Nashville. Paul Allen's background also includes work in political management, radio and television programming and management, radio ownership, and broadcasting work for the US Armed Forces Radio. His consulting clients include companies in the management, public relations, and film industries. He is an alumnus of Leadership Music, and a member of the Country Music Association. He is a recipient of the Department of Recording Industry Outstanding Alumni Award for Service to the Community, and the 2006 Award for Instructional Technology. Paul teaches artist management, Internet for the music business, concert promotion, and marketing of recordings at Middle Tennessee State University. He is author of the new how-to and hands-on book, *Artist Management for the Music Business*, and is co-author of the defining recording industry book entitled *Record Label Marketing*.

Bruce Bartlett

Bruce Bartlett is a microphone engineer and a technical writer for Crown International. He is also a recording engineer and producer for his own 24-track recording studio, offering both studio and on-location recording services. Bruce is a member of Syn Aud Con and the Audio Engineering Society, and has presented several papers at AES conventions. He holds a number of patents on microphone designs. As an audio journalist, Bruce has authored over 900 articles and eight books on audio-related topics. He enjoys music and plays guitar, bass, drums, and keyboard. Bruce received a degree in physics from the College of Wooster, and studied electrical engineering at Gannon College and the University of Akron. Then he worked as a microphone engineer at Astatic Corp. and at Shure, where he worked alongside the designers of the SM57, SM58, and SM81 microphones. He joined Crown in 1982 and helped to develop their line of microphones, including various PZMs, the PCC-160 industry-standard stage floor microphone and the CM-311A industry-standard headworn microphone.

Danny Cope

Danny Cope is Course Leader for the BA (Hons) Popular Music Studies programme at Leeds College of Music, UK and also teaches on the BA (Hons) Music Production course. His teaching specialism is Songwriting, and he teaches modules in Pop Performance, Songwriting, Song Production, Popular Music Composition, and Music Production Project. In addition to his work in education, Danny works as a freelance Bass player, writer, and speaker and has made several TV appearances, and toured extensively around the UK. He has released four highly acclaimed independent CDs as a solo artist and has also written and presented a DVD entitled *Everything You Need To Know About… Setting Up A Bedroom Studio*. As a songwriter he has a publishing contract with DayBreak Music Ltd. Visit www.dannycope.com

Craig Golding

Craig Golding is Course Leader for the BA Music Production course and teaches on both the Music Production and Popular Music degree programs at Leeds College of Music, UK. His teaching specialisms include Audio Recording, Music Production, Sound recording techniques and Studio Musicianship.

Craig has also pursued an active freelance career in Sound Engineering and Production since 2000, including work as a freelance FOH Engineer at The Sage, Gateshead, and Queen Elizabeth Hall, London.

Russ Hepworth-Sawyer

Russ Hepworth-Sawyer is a sound engineer and producer with over 13 years of experience in all things audio and is a member of the Audio Engineering Society and the Music Producer's Guild. Russ is currently Senior Lecturer of Music Production at Leeds College of Music. Additionally, through MOTTOsound (www.mottosound.com), Russ works freelance in the industry as a mastering engineer, writer and consultant. He has also contributed to *Sound On Sound*.

David Miles Huber

David Miles Huber is widely acclaimed in the recording industry as a digital audio consultant, author and guest lecturer on the subject of digital audio and recording technology. As well as being a regular contributing writer for numerous magazines and websites, Dave has written such books as *The MIDI Manual and Professional Microphone Techniques*. His most prominent book *Modern Recording Techniques* (www.modrec.com) is the standard recording industry text worldwide. He also manages the Educational Outreach Program for Syntrillium software (www.syntrillium.com), makers of Cool Edit 2000 and Cool Edit Pro. In addition to all this, he is a professional musician in the downtempo dance genre, and has produced CDs that have sold over the million mark. His latest music and collaborations can be heard on the *51 bpm label* (www.51bpm.com and www.myspace.com/51bpm).

Roey Izhaki

Roey Izhaki has been involved in mixing since the early 1990's. He gives mixing seminars across Europe at various schools and exhibitions and is currently lecturing in the Audio Engineering Department at SAE Institute, London.

Bob Katz

Bob has played the B flat clarinet from the age of 10, and his lifelong love of sound and music led him to become a professional recording, mixing, and mastering engineer (since 1971). Three of his recordings have garnered the Grammy[a] award and many others have been lauded in publications such as *Stereo Review, Audio,* and *Stereophile*. He has written over one hundred articles for audio and computer publications, and is an inventor and manufacturer with processors and support gear in use at mastering studios worldwide. His most recent patent-pending inventions, the K-Stereo and K-Surround Processors, fill a missing link in the mastering and post- production pantheon. He has also been a workshops, facilities and section chairman of the AES and has given lectures in several countries.

Amy Macy

Amy Macy is Assistant Professor in the Department of Recording Industry at MTSU. She received both her undergraduate degree in Music Education and her Master's degree in Business Administration from Belmont University. Amy has worked for many labels including MTM, MCA, Sparrow Records and the RCA Label Group, creating strategic marketing plans for new releases while maintaining a sales and marketing focus at retail level.

William Moylan

William Moylan is Professor of Sound Recording Technology and Music at the University of Massachusetts Lowell, where he has been the Coordinator of the Sound Recording Technology program since arriving in 1983 and served as the Chairperson of the Department of Music from 1998 to 2006. He has been active in Music and Technology communities for over 25 years, with extensive experience and credits as a record producer, recording engineer, composer, author, and educator. He has served on the education committees of the Audio Engineering Society and SPARS (Society of Professional Audio Recording Services), and consulting internationally on program development, assessment, and curricular matters. As a producer and engineer Dr. Moylan has recorded many leading artists and ensembles in a broad section of music genres. His recordings have been released by major and independent record labels and have in wide recognition, including several Grammy Award nominations in a number of different categories.

Preface

Russ Hepworth-Sawyer

When Focal Press approached me to consider pulling some highly reputed authors' work together to make a new publication, I considered what subject matter I thought was absent on the shelves. As a Senior Lecturer in Music Production at Leeds College of Music, it occurred to me that many of the books on our reading lists don't quite map, in detail, what we call the production process from start to finish—from inception to release. This book I hope, to some extent, fulfills this place on the bookshelf.

For the most part, this so-called music "production process," whilst a little fluid in the sense that it can be reinterpreted from time to time, remains loosely the same despite the release format or genre. For today it would be prudent to mention that there are some considerable changes occurring. Recording companies are beginning to consider their tactics in light of different sales and marketing models. These courses of action might have paved a new direction for many future artists and thus upset the "traditional" production process discussed here.

The record production industry is changing nearly as fast as that of the computer industry. As such we recognize the need to keep readers up-to-date when possible. The accompanying website to this book, www.demo2delivery.com, aims to provide additional news, signposts and information about the authors.

PURPOSE

This book should appeal to those musicians wishing to gain a handle on what to expect at every stage of the production process. This book introduces the issues involved and what a musician can do to improve his position armed with the knowledge this provides. It also provides a section called "Doing It Yourself?" for those wishing to take the process in-house.

Although this book speaks primarily to the musician, students of sound recording, music technology, music production, popular music studies and music management should also find this a resourceful guide to the process and their involvement within it.

Each chapter is merely a foundation to a fascinating and vast industry, and does not attempt to cover each subject matter to its fullest extent by any means. Because of this, each chapter is accompanied by a further reading list signposting the reader to excellent resources should you wish to gem up further.

The Editor's Acknowledgments

Firstly I would like to thank my wife Jackie and my son Tom for all their support and patience through yet another project so soon after the last book. (Sorry to be locked away in the studio for so long!) I would like to thank my mother for her support and encouragement that allowed me into music and audio in the first place. Thanks also go to my parents-in-law, Ann and John, for their endless help and support. My thanks and utmost respect goes to Max Wilson for his continued friendship, musical support, tuition and guidance all these years.

Thanks to Catharine Steers for her excellence as an acquisitions editor and for asking me to contribute to the "Brownbag" Project. I would also like to acknowledge the excellent work of all the authors involved with this book. Their contributions here are representative of the excellent and authoritative books they have written for Focal Press.

My respect and appreciation goes to Ben Burrows, Craig Golding and Danny Cope for their detailed and guiding feedback—"Thanks Chaps." Additionally thanks should be extended to all my colleagues, past and present at Leeds College of Music for their support, with a special mention of Barkley McKay (a very knowledgeable chap on all things audio) and Paul Baily (also of Re:Sound UK). I would also like to acknowledge the contribution our students at LCM have made indirectly through discussions over the years. Thanks also go to Iain Hodge and Peter Cook at the London College of Music for their continued friendship and support throughout my career.

Thanks and respect must go to: Tony Platt and Mick Glossop (of the Music Producers Guild) for their contribution to this book; Ray Staff (Legendary Mastering engineer) for making such a generous contribution to the mastering introduction; Andy Barlow (Lamb/LunaSeeds) for discussing topics and quotes on many aspects of this book; Rob Orton (the Hit Mixer) for all the conversations over the years.

**The Editor's contribution to this book is dedicated to the
memory of Christian Watson.**

Introduction

Russ Hepworth-Sawyer

In This Chapter

- Introduction to Processes of Production
- Overview of the Book
  - □ Section A—Preparation
  - □ Section B—The Sessions
  - □ Section C—To Market
  - □ Section D—Doing It Yourself

WHAT DOES THE TERM PRODUCTION MEAN (IN THIS BOOK)?

"Production" can be used to describe so many different things these days. In this book the authors discuss production in two ways.

First is what we would naturally refer to as Music Production, which is an expression of the creative and artistic development of music both in and out of the studio. Whilst this book does not intend to address the creative and artistic elements of music production, it does deal with the back-office work that the producer undertakes.

Second is the Production Process itself. By production, in this latter instance, we refer to the process of producing an end product: the making of the CD; the making of the media which sits on a server for download through an online store; the physical medium perhaps not yet decided upon which will convey our music to the masses in the future, if at all.

This latter reference to production is the process we outline within this book. Each stage of the process has its own innate history and has developed into a well-oiled machine that responds with innovation to change and style throughout time. The snapshot the authors have captured is a concise description of the process as it stands at present. At the end of every chapter, the authors have detailed a further reading. Readers are encouraged to explore these signposts where possible.

INTRODUCTION OF PROCESSES OF PRODUCTION

Anything published, whether to be a book, a magazine, a movie on a DVD, a website or an audio CD, must follow some kind of a so-called production process. In this instance, we do not mean music production per se, but the process of producing a product physically: the making of the CD; the making of the media which sits on a server for download through an online store; a physical medium not yet decided upon which will convey our music to the masses in the future (Blu-Ray, SACD, etc.).

This so-called production process has, for whatever artistic output, been carved out from initial trial and error, and through painstaking refinement to create and define the present systems and procedures we rely upon today. Any artistic process can follow a generic model similar to the one shown below. In this example, we've identified seven rough stages of the production process.

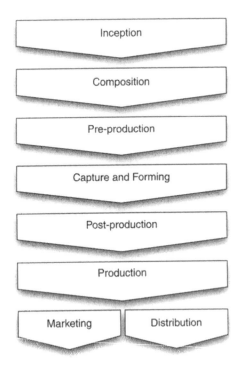

Any creative, yet commercial, process resulting in a final product, can follow a model similar to this. It is worth noting that this is by no means exhaustive and variations can be the norm.

To produce a creative work, an idea must flourish or the desire to capture an experience. So often, many of these ideas fall by the wayside, but from time to time, they will make it to the drawing board. This book begins where the idea of the production process meets the composition in something we have called Song Production. The composition of the work is mainly outside the remit of this book.

The next stage in the process is pre-production. In this book, we dwell for some time on this topic, looking at its importance and some approaches that can be employed to improve productivity in the remainder of the process. Preparation for any creative project can seem to detract from the art form, but in the case of any production, there is a business side that needs to be considered and supported. In this eventuality, it is imperative that planning is given equal value, or equal consideration, when contemplating a project. Is the project viable? How should the project begin? Who should be involved? These are all valid questions that require valid answers before commencement. Answers will inform the rest of the process and the ultimate success, both financially and artistically.

As with any art, it needs to be captured so that it can be portrayed. Paintings need canvas upon which they can be created and later viewed. Music requires some form of "canvas" too. This can of course be manuscript using traditional notation or a sound wave captured in either digital or analog form. The captured work needs to be structured or formed. In music production and recording terms, forming means mixing so the elements can be balanced accordingly. For the painter, this would mean less red, more light, and so on.

Post-production as a stage in the process is the first step on the mass-produced train line. For a painting to be mass produced or copied, it must be encapsulated first. Once copied it can be prepared for mass production. Any edits in light, color or shade can be applied to ensure a maintained quality across the various mediums (postcards, posters, prints, etc.). It is the same for music. Music requires to be prepared for the medium upon which it will be presented or mass produced. Additionally the quality can be improved or balanced at this stage.

Next in the process is the production itself—the way in which the mass-produced material is collated and reproduced. Replication, distribution and marketing are of least interest to the artist compared to the inception, composition and capture of the art. However, it is imperative for the mass-produced reproduction to retain as much quality as possible. As such equal interest should be paid to this part of the process as it is in this book.

OVERVIEW OF THE BOOK

This book has been divided into a number of main sections. Each section is denoted by a group of stages in the production process, other than Section D that introduces discussion about "do it yourself approach."

Section A—Preparation

We begin with Section A: Preparation in which the authors tackle the process of composing the song with production in mind, planning its production process and understanding the roles of the people involved.

Section B—The Sessions

Section B discusses the discrete elements of the studio sessions that make up a production. At no stage are recording techniques dwelled upon as there are plenty of books that deal with this. However, this text looks at preparing you for the process ahead, discussing what to expect from each session of the recording session through to the mastering session.

Section C—To Market

Section C reveals how the recorded material can be brought to market using both physical media such as the CDR and digital downloading. An introduction to the marketing and distribution methods used by labels is discussed.

Section D—Doing It Yourself

Whilst strictly outside the remit of this book, we discuss some items that anyone wishing to manage their production process might wish to know. Again we do not discuss how ProTools works, or how to get the best effect from your Lexicon, but we do discuss things that should help you in a studio session.

With no further ado, let's delve into the process with the inception of the music, the composition and its relation to the process—Song Production.

SECTION A
Preparation

Chapter 1

Song Production: The Marriage Between Composition and Audio Production

Danny Cope

In This Chapter

3

INTRODUCTION

If a song or composition of any kind is ever to be heard away from a live performance, it needs to be captured through some form of recording. By committing a composition to changes in magnetic flux or binary code, its recording will effectively set a composition in stone and represent it on the majority of occasions that it is heard. Regardless of whether you are approaching this book as a potential producer or a composer of some kind, the manner in which the process of production is married to the composition is therefore extremely important. A good marriage takes a bit of give and take and the relationship between a song and its production is no exception.

WHAT IS A SONG?

There are numerous ways of looking at exactly what a song is, especially when we see it in its role as part of a production process. For some it tends to be the starting point that kicks the process off, and for others, it's more intrinsically linked into the production process and emerges and evolves along with the sound and vibe of the track as it is built.

Regardless of the manner in which it appears, a song will always consist of several key ingredients. Whether it be conceived by an acoustic guitar-wielding singer/songwriter, a sampling enthusiast, or a combination of similar approaches, a song will always include elements of form, melody, harmony, rhythm and lyrics. The way these elements meet and marry up can vary considerably from writer to writer and producer to producer, but they will inevitably be required to gel to some degree at some point. If we look to a dictionary or encyclopedia for definitions of "song," we will invariably be presented with a definition such as this one from Grove Music.

> ... a piece of music for voice or voices, whether accompanied or unaccompanied

This is a very basic definition, but it has to be. The simple reason for this is that there are so many different kinds of song. They can be so vastly different that it almost seems unsophisticated to use the same word "song" to describe them all. A children's song, for example, is almost always going to be a huge distance from a Death Metal song with regard to its compositional makeup. Similarly, Hip-Hop and Folk songs tend to see very few things eye to eye, but nonetheless, a "piece of music for voice or voices, whether accompanied or unaccompanied" is classed as a song regardless of the genre that it falls within. The marriage of several key ingredients results in a compositional structure that transcends the class of genre and subculture, and it's these ingredients that give us the most obvious handle when it comes to getting hold of exactly what a song is. However, if we regard a song to be merely a combination of pitches, rhythms and sounds, we are missing something very important in what defines them. A song should not be defined purely by what it *is*. What it *does* is also of immense significance. Before we move on to look at the functional purpose of a song, a brief appreciation of each of its ingredients will be beneficial.

Melody

Regardless of the genre in which it resides, and the degree of effort that went into shaping it, a song's melody is a fundamentally important ingredient. This is because it is usually the part of the song that the listener will be able to get involved in most readily. It is often referred to as the "top line" in a song because it does effectively sit on "top" of most of the other ingredients and is easily accessible regardless of the extent of musical understanding possessed by the listener. Unless the range is way beyond physical limitation, it's likely that someone enjoying any given melody will have a stab at joining in from time to time. It's something that we humans just do. More often than not we feel more comfortable doing it behind closed doors, but getting involved in a song through participation tends to take place when we are enjoying what we are hearing. The melody provides the means to do just that.

It's fair to say that there are plenty of songs that have had a proportionally tiny amount of time spent on the construction of the melody compared to other ingredients. It's also fair to say that this fact has not hindered their success a great deal where the melody was never intended to be the focal point or the reason why the song exists. The "feel," or "vibe," lyrics, or ambient sonic qualities may have been the driving force behind a song's creation, and although these qualities can hold a song up, they cannot completely negate the importance of the position that a melody will inhabit. It may well not be the focal point of composition in all genres, but it will almost always be exposed on the surface nonetheless. Both writers and producers need to be fully aware of this fact. Whether it be a very simple repeating three-note motif or complex labyrinth of inverted and interrupted scales and modes, a melody, no matter how simple or what instrument it is conveyed with, is always going to be a readily accessible element of any song.

Harmony

In case you haven't come across the term in this light before, the word "harmony" here is not being used to describe just what a backing vocalist sings. Harmony is the term used to describe what many songwriters prefer to call "chord progression." The harmony is the chordal structure that supports or harmonizes the melody and that can convey a great deal of information about the genre of a song. Some styles of music such as Country Music tend to be built on relatively simple harmonic structures, whereas others, including Jazz Fusion for example, usually feature much more elaborate use of extensions and chord substitutions.

There are genres where the supporting role of harmony is very evident. It can simply be conveyed through a realistic and transparent recording of an acoustic instrument or three, and generate a believable representation of a live performance. In other genres, however, the harmony can take its place with confidence at center stage. A song's purpose isn't necessarily to be memorable or singable. These are undoubtedly worthy attributes, but making a song good to dance to, or good to chill out to, or good to experience in any other way are also relevant and appropriate qualities for writers and producers to aspire to. A song can exist primarily to showcase a collection of sounds and the harmony will provide the vehicle to convey a lot of that information. It will also provide a platform on which a Producer may find opportunities to pivot towards varying genres. It's not just a case of recording the right notes. Capturing the right sounds is actually a more pertinent pursuit in some cases. Writers and producers should never lose sight of the purpose of a song, and the way in which the harmony is written and recorded will do a lot to define a song's character and intention.

Lyrics

From the above definition, it is clear that a song needs to include some words to be sung in conjunction with the melody. Without lyrics, the melody in a song ceases to be part of a song, and becomes just a melody. This would render it more of a "composition" than a "song." Similarly, a lyric without a melody behind it ceases to be part of a song and becomes just a collection of words or prose.

When lyrics are written, they can define genre. At the very least, they can suggest them. There are some themes that are universal and can be found in all genres and styles of production. Songs about love can take many guises and can be found nestling in all sorts of different styles of music for example. However, there are some lyrical themes that will not be particularly welcomed or appreciated in some genres. When a song is written and produced, there needs to be a considered understanding of what it communicates and the manner in which it is communicated. If lyrics can define or limit genre, then they can also do a lot to define production style.

Rhythm

The previous three ingredients all occur concurrently and are in effect stacked on top of each other in the creative process. Rhythm doesn't quite fit into that club. Owing to our entrapment in the space–time continuum, rhythmical elements all take place after each other, and the timing of each event is just as important as each event itself. The speed of the song, the regularity of the harmonic pulse, variation in stress pattern, and the changes in chord and note length are all incredibly important in defining whether a collection of neatly stacked notes will be magical or plain old monotonous. Time signatures are also important tools in the rhythmical makeup of any given song. Like harmony, the manner in which a rhythm is presented affords the producer the opportunity to land a clear production stamp on the recording. Whether it be through the complexity of the arrangement itself such as the interplay of polyrhythms, or the textures and sounds used to create them, the presentation of a rhythm will have significant impact on the perceived character of the song it is embedded in. Even the most unimaginative sequence of chords, words and tones can become interesting when they are presented in an unusual and unfamiliar manner, and it is this manner that we will get to shortly.

Form

In addition to rhythmical elements, the other linear-based creative decision rests in the construction of the form of the song. The writing of songs is generally categorized through naming each section. Pretty much any song we come across will contain a chorus or pay-off line of some kind, and this moment will be surrounded and usually held up by supporting sections such as introductions, verse, bridges and alike. Editing the form of a song is often massively important in making a song "work." A song can consist of all the right ingredients, but if they are thrown in the pot in the wrong order, then the recipe just won't work and the result will be inedible. Cliché phrases such as "Don't bore us, get to the chorus" are often heard and with good reason. The stage of ordering the different sections of a song is where we can carefully present the information as a purposeful chain of events. Things that are generally important to consider in the structuring of a song are the need to make sure it is an appropriate length for its target audience and market, that it contains information that is interesting with as little padding as humanly possible, and that it achieves what the audience wants to hear with regard to listening complexity. Some audiences like songs that consist of ten or more sections. Some don't and would rather have a very safe and predictable passage through the song to a tidy conclusion. Whichever way a song goes about its business, the order in which the different sections appear, and the manner in which they slot together is very important. Reflective editing is in effect the only thing that separates song writing from improvisation, and should therefore be taken very seriously.

Opportunities for the editing of song structure present themselves consistently throughout the production process and should always be confronted head on when they arise. Generally speaking, it's a good idea to ensure that the first draft of any song has the basis of the song all there so that editing can be a process that hones rather than adds. It may be that the composer or composers choose to edit the song before it reaches a producer if the compositional approach makes that a possibility, but even in those cases, the song should still not be above further editing as the process progresses through each stage. Getting the song to fit where it is heading will almost always require some chipping away at the edges that are stopping the substance of the song fitting into the hole it's designed for. Some of these protruding edges are more obvious than others, and it may be a while into the production process before it becomes apparent they are an issue that needs addressing. One of the most valuable things that a producer can bring to a song is objectivity, and it's often in the structural makeup of a song that a producer can make the quickest and most effective changes.

Like a good partner in any marriage, a producer will be able to help a song build in confidence and identify and pronounce good and admirable qualities within in. A good producer will be honest with the composer and composition and help extract the best out of it. This process in itself will invariably lead to hidden gems in the composition coming to the surface.

IS A SONG JUST A COMBINATION OF ITS INGREDIENTS?

So, what can we do with these ingredients? The main task at hand is to ensure that they all slot together neatly and enhance each other to maximum effect. If we were to visualize the combination of ingredients, it could be easy to see them as being stacked on top of each other. This would certainly seem to work as far as the writing process is concerned, where the melody *may* have been written first, followed

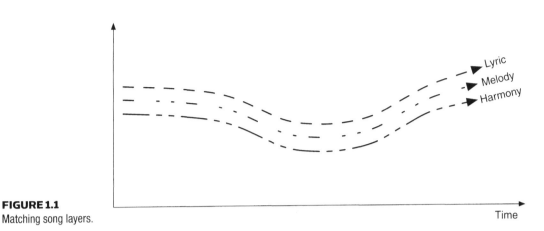

FIGURE 1.1
Matching song layers.

by a harmonic chord progression to underpin it and then followed up with some lyrics that work with the melody.

Figure 1.1 is obviously a simplified representation of the manner in which these compositional elements fit together, but it serves its purposes for the time being. The reality is that every writer, regardless of experience or skill, will have a collection of snippets of songs lying around the place. It's likely that there will be the odd few lines of lyrics scribbled down on scraps of paper, melodic phrases sung to death but as yet unfinished, chord sequences that sound "nice" but inconclusive, and possibly a few "vibey" tracks stored away on a tape or hard drive somewhere just waiting to be let loose. Coming up with the occasional idea here and there isn't generally that difficult. What can drive us slowly insane is figuring out where each of these individual moments of brilliance can find their life partners and make truly beautiful music together. The first thing to remember here is that the writer should never rush this process. If two ideas don't go together in their current form, then they don't go together in their current form and that should be the end of it. If a writer is crafting a song on their own, it can be very easy to lose an objective view of the song as it evolves. The goal can get clouded and the composer can get so close to the composition that they really can't see what they have in front of them anymore. One of the benefits of the production process is that it will invariably introduce at least one more pair of ears to the process at some stage. In addition to facilitating the recording and presentation of the song, the person filling these shoes, often the producer, will be able to offer this objective opinion and help keep the crafting process focused and beneficial. Sadly, it's a common problem to see writers who are struggling through on their own marry off incompatible ideas simply so that they can just get shot of them. That will never do anyone any favors other than possibly a short-term favor for the writer, but even that is likely to come back and haunt them (Figure 1.2).

The ingredients we have to play with need to be nurtured and slotted together neatly. On occasions, they will fit together nice and snug. This can happen naturally as a result of writing different elements concurrently such as the melody with

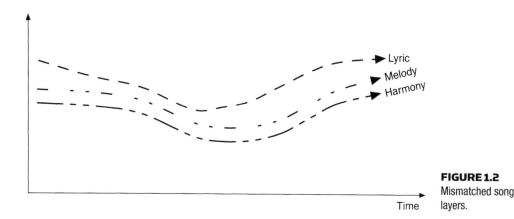

FIGURE 1.2
Mismatched song layers.

Time

the lyric or harmony with the melody, and sometimes the writer can be fortunate enough to have two or more separately fashioned ideas that seem to have been made for each other fall into each other's arms after the event. On other occasions, however, they may not go neatly together in the first instance but will be able to work something out through a little bit of give and take. This may require a lyric to lose a word or syllable here and there, and the rhythm of the melody may have to adjust its run a little bit, but this often happens in the process of writing anyway and is no bad thing.

Different Measurements

Just as different recipes and tastes require different quantities of each ingredient in food, so too do different genres require different weights of each of these ingredients. The word "weight" has been used here as opposed to "amount" for a simple reason. If a song is 3 minutes long, then there will be 3 minutes of each ingredient present for the most part. There may be moments where the vocal melody departs for a moment along with the lyrics but there will be constant rhythmic and harmonic information at play throughout. The weight is about how important each of these ingredients are at any given time in making the song work. Some genres of music are typically more dependent on some ingredients above others. For example, folk songs tend to rely very heavily on lyrical content because they commonly exist primarily as a platform for the lyric to say something. There may be an age-old story that is being conveyed, supported subtly and sensitively through a simple chord progression and arrangement. In many songs such as this, the harmony is nothing to write home about because it isn't supposed to be the focal point. It can just support and play its part accordingly. If it suddenly got really interesting and all sorts of inversions and extensions were at play, then there is a danger that it would start to fight with the lyrics at the front of the stage and distract rather than compliment. Similarly, there are many "Urban" songs featuring an interplay of rhythmic elements in the production that are more of a focal point than the authenticity and historical grounding of the lyrical narrative. In these cases, the "vibe" of the track is often what the song is really all about, and the lyrics are there more as a means to an end rather than a focal point in themselves.

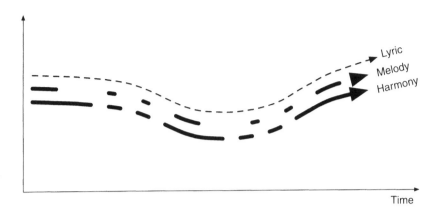

FIGURE 1.3
Where lyrics are less of a focal point than other layers.

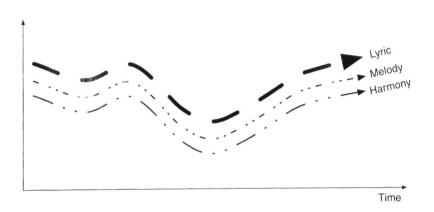

FIGURE 1.4
Where lyrics are the focal point.

The subtle variation in the balance of ingredients needs to be considered carefully, and even though decisions regarding the arrangement and sonic production will frequently come after it has been written, it is still wise to have these considerations in mind when crafting the song in the first place. Some songs will have a fairly even spread of weight right across the ingredients and some will have a very obvious tipping of the balance employed. The significance of each of those ingredients varies according to what the song is trying to achieve, and to a large extent, the genre that it is found within.

What Does the Song Do?

In addition to looking at the ingredients that go into making up a song, we also need to look at what the song does. Understanding how a song is put together does not necessarily mean that we understand what it is doing or how it does it. We know a lot about the human body these days, but there are still millions of things that we don't quite understand and that scientific analysis is never likely to answer. For example, we understand how the ear works in principle, but can't quite seem to get a handle on why different combinations of waveforms hitting them will generate different emotional responses. Beyond the physical makeup are the emotional and spiritual elements, and this principle applies to songs as well.

We can achieve all our musical grades and get 100% on any music test we ever take, but that will not necessarily make us better songwriters. It means that we know how to control our tools and that we can craft effectively, but it doesn't mean that we have the skill or understanding required to manipulate emotional response in listeners. There are plenty of songs that seem to do everything "right " with regard to compositional makeup, but that leave the listener stone cold. On occasions such as this, it may well be that a producer can make all the difference. In any team creating a recorded product, it can be hugely beneficial to have someone who isn't musically educated. It can be easy to forget that the majority of the listening public will relate to a song on more of an emotive level than an analytical one, and having someone in the team who can offer an "outsider's" perspective in this respect can be massively helpful.

It doesn't matter who that person is or what role they fill. A songwriter with no formal musical training write a song without really understanding it and have a producer focus it a little through applying some core musical principles, or a producer without a traditional musical education can help convert a "correctly" fashioned song into one that goes beyond being competent and into extraordinary.

A song is more than the sum of its parts. In fact, it could be argued that the song exists before any of the ingredients are wrapped around it at all. The ingredients are the tools that enable the writer to give the guts of the song a physical representation. Musical ingredients are essentially an outworking of mathematical principles, and if songwriting was just about combining them, they might be interesting and clever, but they would never really be any more than that. The soul or guts of a song is what can make it magical and wondrous. The ingredients we use to give body to songs are just tools that carry and embody, nothing more and nothing less.

The chapters that follow in this book are all very useful in playing a part in sharpening our tool sets and enabling us to get a handle on what we can do in the process of getting a song from demo to delivery, but it is fundamentally important to grasp the fact that if the guts of a song aren't in place from the start, the rest of the process will be nothing more than artistic and scientific endeavor. It's all about the song and the vast array of options open to the writer and producer can be so overwhelming that it is easy to forget this point sometimes!

WHAT'S THE POINT OF WRITING A SONG?

In seeking to ensure that the quality of the song itself doesn't get neglected under the pile of instruments, plug-ins, monitors and master copies, it's a good idea to start the writing process with the question "Why is this song being written?" It may seem like a mind numbingly obvious thing to ask, but the reality is that this question is often bypassed entirely so that the fun of playing with buttons and automation can begin. The simple truth is that if we don't know why we are writing, the chances are that we won't do a very good job of writing the song in the first place. Even the most basic thoughts such as "What is the song about?" often go unanswered. Nine times out of ten, the reason why writers struggle to finish songs is because they don't actually know what they are writing about. The song has kick-started with a flash of inspiration and got off to a flyer, but that inspiration has not turned into perspiration and has remained the fuel for about 20–30 seconds of promise. There are obviously exceptions to this where songs just seem to appear through just getting on with it, but for every one that works like that, there tends to be a considerable number more that grind quickly to a halt. Knowing the point of a song is massively beneficial. Firstly, it gives us something to work toward and therefore aids the focus and creativity in the writing and production. Secondly, it gives us some kind of meter by which we can assess how successful we have been in achieving our aims.

What Is It Supposed to Achieve?

We can know what it is about and who it is for without knowing this piece of information. Knowing what the song is intended to achieve enables the writer to be considerably more focused in ensuring that the desired outcome becomes a reality. It will also help set the goal posts for the entire team who will be working on making the product fit for purpose throughout the entire production process. Whether it be to generate an emotional response, to advertise an event, to accompany an occasion, or simply to get something off your chest, a good song will serve some kind of purpose in addition to just existing. The "what," "who" and "where" are all important links between what the song is and the manner in which it will be presented.

HOW DO WE IDENTIFY WHO THE SONG IS FOR?

This book is all about the process through which a song will make its way through the necessary steps to arrive at someone's ears dressed as *suitably* as humanly possible. Knowing where the song will be "delivered" is useful information to know so as to ensure that the content of the package is right for the recipient. There's no point sending someone something that they don't like. The creator gets to choose whether knowledge of the recipient dictates the creation and treatment of the song, or whether the song's creation and treatment dictates who the likely recipient will be. Songs will appeal to different people on different levels. For some, the compositional makeup of the song is the most important part. For these, the interplay of rhythm and harmony, and of lyric and melody will be where the appeal is grounded. It should be noted here, however, that there are plenty of recipients of songs that will not be attracted on these grounds at all. For many, the attraction to a song is generated not through its content, but the manner in which it is dressed up. The production process is what does this dressing and advertises the song to its full potential. Even if the recipient is able to understand how cleverly crafted a song may be, there is no guarantee that this knowledge will help them get past several other stumbling blocks. One such stumbling block is the production style that the song is wrapped in. When songs are written, the genre that they will fall into can be very obvious even when the song is in its most raw and delicate state. The lyrical theme or groove for example can dictate whether a song is likely to be more suitable for a folk or reggae market straight off the bat. However, it is possible to write a song that facilitates numerous possible routes into all sorts of different genres owing to the fact that its content isn't pushing it anywhere specific. The genre a song lands in isn't always dictated by the song itself. Sometimes it needs to be decided by the producer or arranger at a date after it has appeared in its first form. In fact, there are occasions where a producer may be so sure that the song needs to head to a certain audience that elements of the compositional make up are altered to ensure that its journey into that marketplace is more focused and effective. This marketplace is very important when it comes to viewing the song as a "product" which is essentially what it is when it comes to getting it "out there" to people. Regardless off the impassioned heartache and personal soul-searching that has gone into making a song authentic and "real," the moment it starts its journey into the big wide world, it becomes a product. For some reason, there are plenty who frown on people "selling out" in this way. It's not "selling out," it's giving out.

The manner in which a song is dressed up and sent out can have a direct effect on whether people will like it at all. When music fits into different genres, which it almost always seems to do to aid marketing and clarification, it also falls into the grip of the subcultures that surround it. When a song and its production gets classed as "Death Metal" for example, it finds its way into the "Death Metal" camp and into the presence of the attitudes and dress codes, etc., that typify that genre. What this then results in is the fact that unwritten rules of taste dictated through fashion and trend above musical appreciation come into play in deciding how accepted that material is going to be. If a song and accompanying production from a foreign camp are thrown into the mix, then there is a real threat that the production itself will get cast aside without a second's thought. The reason for this is that a large number of recipients in the marketplace listen to what they think they are supposed to be listening to above what they can learn to appreciate on a musical level. It happens right across the board in every music genre. Painful though it is to have to acknowledge, there are plenty of recipients who will immediately dismiss a song simply because they can't associate it with how they view themselves and the subculture and accompanying genre or genres of music inherent in that identity. So even if the song is superbly crafted and sent out to breathe with the best upbringing possible behind it, the recipient may cast it aside without a second thought for reasons that are nothing to do with the song at all. Marketing, introduced later in this book plays a massive part in getting the songs to those who will appreciate them, and also in telling them why they will appreciate them. There's a lot more to it than simply combining ingredients as cunningly and effectively as possible. In addition to facilitating a more focused writing and production process, knowing the vision for the product will also enable its marketing and publicity to be more tailor made. If the creator goes with the song first recipient later approach, it could well be that the song is fashioned in such a way that makes it completely undeliverable. As touched upon above, there is no guarantee that the audience will want to hear the song at all if their position has not been considered in its construction. The creators may get lucky and fashion something that fits nicely into a market without much prior consideration, but from a production perspective, it makes much more sense to have the needs and wants of a designated marketplace in mind before or at least at the early stages of a song production.

If knowledge of a target market is useful in focusing the production process, it follows that it makes sense to purposefully glean as much useful information from that target market in ensuring the resulting product ticks all the boxes it needs to.

What Works?

An obvious place to start in discovering what will work for a designated audience or recipient is just to ask them. Obviously it can't work as simply as this as there will be too many people to ask, and asking them would not provide useful answers as the majority of recipients would not be able to provide the kind of concise musical answers we would require anyway. What we can do is to look at what else that market is listening to. Through identifying the kinds of songs and productions that are popular among the audience that we choose to target, we should be able to get some kind of idea what they will like to hear.

It is interesting to note that target markets tend to identify and present themselves quite coherently whether they mean to or not. It's common for fans of certain artists and their body of songs and productions to share a common interest in other similar artists. Website retailers have got wise to this fact and use it to their advantage in purposefully targeting products at browsers who they have good reason to believe will like what they are pushing. www.amazon.com and www.play.com are just two of many good examples of such sites. Phrases such as "shoppers who bought this also bought …," and "like this?, then you'll probably like this too …" are commonplace and are actually very useful for both the retailer and the buyer. The retailer can count on it as an effective marketing tool to target the most likely demographic to part with some cash, and the buyer can use it as a means to discover new artists that they will probably appreciate discovering. It's not just retailers either. There are plenty of other places where this mentality can be seen at work. Online radio stations are another example where the online community of listeners can introduce other listeners with an identified similar taste to new music whether they do it intentionally or not. The website www.last.fm is a good example of this, and nowhere is it more obvious than at www.audiomap.tuneglue.net. Knowing what people are listening to can be more revealing and more helpful than actually asking them what they want to hear.

Through what they refer to as "Hit Song Science" a company called "Polyphonic HMI" has developed a scientific means of determining the " hit " potential in any song they choose to look at. Through a process they term "Spectral Deconvolution," they have systems in place to analytically assess more than 60 parameters of any given song, and to assess how viable the combination of melody, harmony, rhythm, tempo, octave, fullness of sound, pitch, brilliance, etc., are in giving the song "hit" potential. This may seem a little far-fetched or painfully cold and calculated, or possibly a combination of both, but the principle is actually very sensible and clever. The basic premise behind the technique is to compare the results of the analysis to other songs that have gone on to be hits before. Polyphonic HMI have developed their own "Music Universe" in which they have plotted previous hits like little stars in the cosmos. These songs have been plotted in accordance with the results that came out of their analytical assessment, and it shouldn't be surprising to discover that a lot of the songs assessed actually cluster together in different parts of this "universe." This clustering together reveals that their musical makeup and production sound actually shares a lot in common. We have just acknowledged that this similarity exists in the "real world" where fans of one artist tend to share an interest in other similar artists too, and "Hit Song Science" provides a scientific framework through which the level of this similarity can be quantified.

One of the main sources of inspiration for a budding songwriter is to hear a song and to think something along the lines of "that's a great song and I'm going to write one just like it." The writer then has the option of carefully analyzing the song and trying to replicate something similar as a sort of pastiche, or just going with a similar "vibe" and hoping the result will be pleasing. Generally, the second of these approaches is considered to be more credible as it seems less like cheating, but that isn't necessarily the best way forward. The scientific approach that Polyphonic HMI takes to the assessment of these songs is exactly the same the songwriter will have, it's just that the guesswork is removed and some hard evidence takes its place. Obviously, there is more than just the quality of song and production itself that goes into making a song a "hit" and Polyphonic HMI are keen to state that even a song they assess as having massive hit potential won't necessarily go on to be one. Unless it is marketed appropriately and effectively, a song production will not get the chance to be heard and appreciated regardless of how magnificently constructed it is. Writing a cracking song and making it sound incredible is just a part of the mountain of work that needs to be done.

Hit Song Science works its scientific magic at the end of the production process. That's all well and good in giving a result once the product has been completed, but what about at the start of the process? Is there something that the writer can do to ensure that the product will start off on the right foot? Emulation would seem to be the most straightforward approach to employ at this stage. Providing alternate versions and adaptations of previously successful models is always a sensible business strategy. If a product seems to be doing well, then it makes sense to "jump on the bandwagon" and to put out something similar. It is certainly the sensible thing to do in the business marketplace where projected sales and forecasts are as close as we can get to predictions of how things will go. It happens in music too. It doesn't take long to think of a song and accompanying artist that came along with a massively successful mold-breaking song, only to be followed by a small but prominent group of artists and/or bands that provided something similar. This happens because those controlling the purse strings at the labels need to make some money, and they are able to do so through following the trailblazer at the front once they have established a new product and paved the way for some new growth in the market. It can seem very unmusical, and that's because it is. At a relatively early stage in the lifespan of a song production, it ceases to be a 'song' to those who market it and becomes a "product" or a "unit" much like any other.

If this pastiche and jumping on the bandwagon happens so often, how is it that new genres and new types of production get a foothold in the market? It tends to be where money isn't a driving factor. The majority of labels endeavoring to survive in the market simply can't afford to take many risks with regard to trying something new. The Internet revolution has hit record labels particularly hard and has resulted in the scaling back of operations, decreased advances and alike. Those in charge have staff to employ and salaries to pay and therefore need at least some products that are almost guaranteed to sell. Whether the chief executive happens to like the product in question personally is rarely of any consequence at all. It's generally where money gives way to a sheer love of creativity and something new that new songs and productions can be given space to flourish. Once that new sound has had some time to develop and invariably grows with popularity, it's then that it can start to turn into more of a commercial viability for some of the big labels. It is at this stage the "underground" nature of the new genres tends to be brought "above-ground" and marketed to the masses. The development and mass marketing of House Music is just one of many good examples of a relatively small and local music phenomenon being taken aboveground and transformed into a mass produced product.

It's not all about creativity free of financial shackles though. There are also times where bands and artists with a massive and passion ate following are able to release something completely new and "fresh" relatively safe in the knowledge that a large percentage of their fan base will buy it because it's them and not necessarily because they like the new material. Almost everybody has at least an artist or band or two that they have taken to heart, and whose new album they will always buy. That level of loyalty can lead to new ideas, songs and sounds making their way safely into the public arena too.

What then can the writer and production team do with all of this information? If a song production is going to do a good job of appealing to a target audience, then it needs that target to be identified. It also needs to tread carefully between providing something that the designated audience will want to hear and something new enough to warrant them paying some attention. If all new productions were simply reworks of previous successes, then the musical marketplace would quickly drill itself into the ground through a gradual and gut wrenching decline in variety and adventurous output. A good product will be one that is similar enough to be safe, but at the same time, different enough to warrant it's existence in its own right. A way of working within this understanding is to purposefully create what I term the "Triangle of Influence" (Figure 1.5). It must be stressed here that this is not the only way forward. It is, however, a useful technique for exploring possibilities defined by whatever the current music market place happens to be revealing.

The premise is very simple. If we can acknowledge that artists who share a similar sound and composition style also share a similar fan base, we can use that knowledge to intentionally round up and satisfy that fan base through giving them what they want to hear. The intention of the "Triangle of Influence" is to identify three key ingredients in the song productions that are succeeding in the marketplace, and to intentionally place each of them in different corners of one triangle. The idea is that the writer/producer

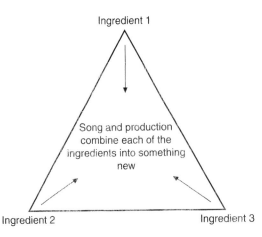

FIGURE 1.5
The "Triangle of Influence."

©2009 Danny Cope

can then create a track that sits in the middle of the triangle through borrowing elements of the productions that surround it. What is particularly important about this approach is that the elements plotted in each corner need to be carefully selected on the grounds of their appeal. The writer/producer needs to select corners that will draw the desired fan base, and this is done through carefully and purposefully identifying the elements of production sounds that are key in attracting the audience that are drawn to them (Figure 1.6). It's a technique designed to sharpen the writer's and producer's appreciation of current market trends.

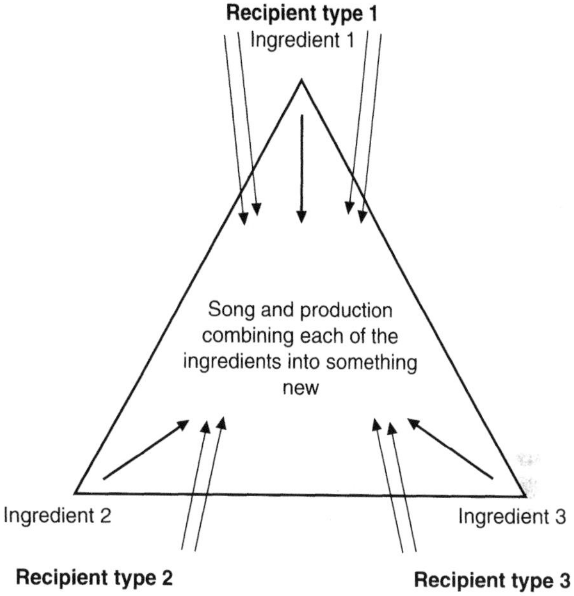

As already identified, a fan base will cluster itself together around several similar artists in all sorts of different genres and corners of the marketplace. In doing this, they can reveal to us which artists share a common thread, and then it's just a case of pulling out the common musical denominators from each of those artists. Factors contributing to this clustering together

FIGURE 1.6
Caption Not Provided

can be any one or a combination of a vast array of potential ingredients, but it is generally fairly obvious what the key components to the "flavor" are. It may be just one important element such as the fact that the productions tend to feature guitar solos, or they are built around solo male singer/songwriters for example. However, it may be that the appeal is grounded in a much more subtle and intricate combination of ingredients such as a combination of the technical proficiency evident in the playing of instruments, the lyrical theme of the material and the structural form of the material performed. It may be that just one of those elements on its own is of no particular interest to the fan base in question, but that the combination of those elements is what really works for that audience. These are the questions to be asked by the writer/producer and the answers should help populate the corners of the triangle. These corners should then be able to act as tempters in drawing fan bases at least to the edge of the triangle.

Once the production has lured the listener to the edge through sounding similar to existing material, the idea is that the song production created through borrowing elements from each corner around it will pull the listener into a slightly new sound. The sound should be relatively comfortable as it will include elements that the listener will already be familiar with, but it should also provide some additional interest from another corner manifesting elements of song production as yet undiscovered or unappreciated (Figure 1.7).

When the triangle is working properly, it should be attracting and bringing together audiences who may as yet have not been un-introduced. The reality of the model in practice is that the majority of each separate fan base will linger closer to the

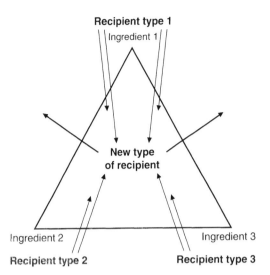

FIGURE 1.7
Caption Not Provided

familiar corner than others. That is to say that their interest in the new song or production will rest mostly in it's similarity to what they are comfortable/familiar with, but the added flavor of the other genres gives the taste just that little bit of an extra something. There are, however, some of the audience who will fall completely for the new sound they are hearing and will stand in wonder right at the center of the triangle having been lured completely by the combination of elements hitherto unexperienced in unison. Once these recipient types have been lured by the new sound, they will also be introduced to each other's taste in music and may well find themselves moving on to discover new things as a result. These new things may be more of a voyage toward the other two corners of the triangle, or out of the triangle altogether to discover another sound somewhere else armed with their new discovery.

There are a multitude of different ingredients that can be placed at each corner of the triangle. Some will be inherent in the musical composition such as those discussed earlier. The shape of the melodic lines employed or the lyrical themes may well in themselves be sufficient to draw a crowd. Similarly, it may be instrumental factors such as the employment of acoustic guitars, fiddles, bagpipes, steel drums or Hawaiian nose flutes, etc., that can lure the listener. There are plenty of music fans whose music collection could be classified predominantly through instrumental lineup inherent within it. There are also the production factors to be considered, such as the quality of the sheen or mix, the use of effects in the production and authenticity or "realness" of the reproduction of the instruments recorded or simulated. The list could go on for pages and pages. Ultimately, it doesn't really matter what is allowed in each of the corners as long as it can be seen to draw a crowd. Whatever is placed there should act as bait attracting prey to the new things laid out before it. Once the prey (audience) have pulled up to the triangle, it's up to the other triangles to do their bit in luring them into the center to experience the full flavor.

Of course, the triangle of influence doesn't have to be a triangle, although changing it would mean having to change the name a bit. A similar result can be achieved through plotting a new idea between just two sources, or it can be a pentagon or dodecahedron, etc. It makes sense to keep the number of influences to a minimum though, to ensure the combination of flavors isn't so complex that it doesn't taste disgusting, or of anything in particular. If constructed and implemented effectively, a triangle of influence will give the audience something that they didn't know they wanted—until they got it. If a song production can achieve that, then it has definitely got something right!

CONCLUSION

In combining the pursuit of writing a great song with the production tools that shall be discussed in some detail in the coming pages of this book, the writer and producer both have to be careful that the process doesn't override the purpose. The reason for a song's existence should never be lost amidst the technology that captures it.

It is possible for a writer or producer to get so embroiled in the pursuit of the elusive vibe and kicking rhythm track, or something similar, that the point of the song gets completely lost.

Where this happens, there is a real danger that the resulting production will be little more than something that sounds interesting but has no real substance. It can be an interesting shell that has nothing in it once it has been cracked open. It's a sad truth that there are a lot of song productions that do exactly this. They sound great for a season, but when the current trends that they adhere to die, they are little more than collections of dated sounds. Only the most significant and seminal tracks will stand the test of time, and that will often be more as museum pieces than anything else.

The song has to come first in priority at least, if not in the compositional chronology, and this warrants the highest level of attention throughout the entire production process. The writer/producer should always evaluate the mileage in any given song production. Sometimes it is entirely fitting to produce something in the knowledge that it has a limited lifespan, but it's good to look beyond that wherever possible.

Traditional song writing ingredients can be divided up differently according to genre, and production elements need to be treated the same way. For some genres, it's obvious where DAWs, Samplers and Sequencers have evidently been huge contributors to the compositional makeup of songs. Pretty much any form of "Electronic "music will have had to use these devices through the very nature of the genre for example. Hip-Hop and House, Trance and Drum'n'Bass are other obvious examples.

There will be fans of some songs who claim that the production isn't a factor in their appreciation of the song as it may simply be a piano and vocal arrangement with no trickery at play in its recording. What needs to be remembered here is that there actually has been a considered production style employed. The fact that it might not sound particularly glossy and that hundreds of plug-ins haven't been applied doesn't mean that the track isn't produced, it just means that it has been produced to sound more natural than it could otherwise have been. That in itself is no easy feat!

In seeking to marry a song with a production style, it's important that the producer understands the genre and the audience attached to it. They need to understand elements of that genre with regard to all its ingredients whether they be compositional or sonic in nature. It is entirely possible to have a good production of a bad song and a bad production of a good song. Where it isn't due to engineering incompetence, the former will usually be where a well-crafted song has been railroaded into the wrong genre because the producer didn't understand the song in the first place. The latter will be where the producer has a great deal of skill and can make even the dullest song sound interesting through a mastery of arrangement and engineering. There are a lot of songs that sound great in their produced state, but sound just plain tedious when stripped of the production that dresses them up. Whether this is a perceived problem or not depends entirely on the genre, as some songs are just not meant to be undressed. They exist primarily for the clothes on their back to have a platform to shine. Some, however, are created to sound great regardless of whether they are performed solo on any instrument, or whether a 20 piece band are giving it all they've got.

It all comes down to purpose. A good writer and producer should know exactly what they are trying to achieve before they complete the process. The vision needs to be in place and the song itself should tick all the appropriate boxes, even if it's just to act as a structure to dress up with arrangement, groove and vibe. A writer should be able to take pride in what they are writing and satisfaction out of how successfully it lands at its target regardless of personal taste. Getting a product from demo to delivery depends largely on starting the project facing the right direction. Once the goal is in sight, the tools are in place, and the demands of that target audience are appreciated, it's simply a case of getting the ball rolling....

FURTHER READING

Cope, D. (Forthcoming). *Righting the Wrongs in Writing Songs*. TCT Press.
Pattison, P. (2001). *Writing Better Lyrics*. Writer's Digest Books.
Perricone, J. (2000). *Melody in Songwriting*. International Music Publications.

Chapter 2
A. Pre-production

Russ Hepworth-Sawyer

21

Planning is essential whatever the nature of the project. Many things vary but all have a common purpose, it's only the way they appear in a project that differs. The following chapter contains things you should definitely consider before embarking on a production.

Ben Burrows, Academic and Head of Production, Mithril Productions

INTRODUCTION—WHAT IS PRE-PRODUCTION?

Pre-production as a name simply describes what should happen before a "production" starts, whether that is before shooting a film or preparing for a West End play. Following on from this comes "production" which in music terms is considered to be the recording and mixing sessions clubbed together, despite often being discrete processes in their own right. Once the mix is achieved, it is moved on to post-production, which in terms of the music industry means editing and mastering.

As a term, pre-production is often a misunderstood part of the process of making music. Misunderstood because many of its facets are not always easily described. For example, pre-production can represent the preparation for a recording session, including thorough rehearsals that could include some pretty advanced recording that may or may not appear on the final mix. Alternatively for electronic artists, pre-production and production can, to some extent, merge into each other as we'll discuss later.

Arguably, the process is defined from the moment a project is devised and comes into existence. In Chapter 1, Danny Cope spoke of the inter-relationship that now exists between song composition and music production. Those decisions, or eventualities, may have immediately led to certain pre-production elements being applied that may need to carry on through to the final record, whether it be a specific sound of a unique synthesizer plug-in, or the heavier use of delay than normal on the vocals upon which the whole track is now sonically hinged. In some cases, it is the pre-production stage that will inform the performer or production team whether a composition, or an idea within it, will or will not work and hint at the developments that still need consideration.

Throughout this chapter, we refer to a "traditional production process," which signifies a model and nothing more of how we sometimes perceived the stages to be: clean cut and definitive. Perhaps once upon a time, these stages were devised partly necessity and partly defined by the composition process itself (Figure 2.1).

Compositions in the pop and rock genres were typically based on a piano or a guitar and subsequently etched into memory or were translated on to manuscript paper to be realized at a later date. The composition would then have been rehearsed with a band, or the manuscript passed on to hired musicians at the start of the recording session. By the simple virtue of this arrangement, the divisions between the stages were much clearer. The recording session was expensive. Expensive because of the sheer cost of the equipment housed or even developed by

The traditional production process overview

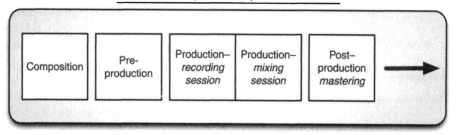

A potential modern production process overview

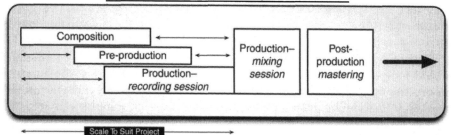

FIGURE 2.1

The Process of Production. The "traditional" model of the production process shows defined stages with no blurring. The latter model demonstrates the fluidity of the early processes in production. In this latter example, pre-production can blur backwards into the composition stage and equally draws back some of the traditional roles from the recording stage.

the studio in question; the range of personnel required for the session from hired musicians, engineers, tape operators through to arrangers and the producer.

Recording sessions in the early days were often restricted to 3 hours as set out by the unions and it was assumed that the ideal performance could be captured within this time. Given this, the importance attributed to the preparation and rehearsals that make up what we can refer to as pre-production was paramount. An example of this was one of The Beatles' first sessions on the 4th September 1962 between 7:00 and 10:00 p.m. in which no less than 15 plus takes were recorded of Love Me Do and a number of takes of How Do You Do It. Within these 3 hours mono mixes were made of each of the pieces (Lewisohn, 1989). This length of session could have been a "demonstration" session for management or could quite easily be the recording session.

The Demo

The demonstration session, or the creation of a "demo," has been an important part of the production process over the years and has offered a tangible calling card for getting gigs as well as wooing prospective producers and engineers to work with. The demo should therefore be part of many a pre-production process as it also gives a recording for the band to listen to and reflect upon in order to re-develop their songs, more of which we'll discuss later (Figure 2.2).

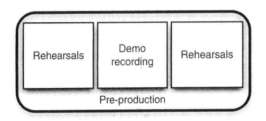

Many producers are said to consider the traditional "demo" dead for many applications. For example, established artists may feel the demo is perhaps no longer necessary as they're working in their professional standard home studios; therefore, to some, the audio may be considered part of the early sessions. It is presumed these original ideas may contribute to the final mix.

However, the demo is very much still alive and kicking for those bands not yet established or newly formed, as these will be the "business cards" for the band. The recording of demos is now so much easier to achieve using affordable, professional standard equipment. This availability of equipment has spawned a whole generation of home recordists who can provide that demo service. How these recordings are made are outside the main remit of this book, although some aspects are introduced in Section D.

How these demos are used can vary. Most material is now data compressed to MP3 and posted on sites such as MySpace which has proved very successful in sharing an artist's music to the world. The demo itself has arguably expanded from solely a physical entity to that of a multimedia space including video links and audio.

Is Pre-production Necessary?

Up to about thirty years ago, the attitude to making records had developed and changed to such an extent that an artist might have had the privileged opportunity to write, rehearse, record and mix all in the same expensive studio. The financial advance from the label and potential sales were adequate to cover this in addition to all the fabled lifestyle excesses we're familiar hearing about. Attention to details such as planning, rehearsing and arrangements for session musicians seems a far cry from this hedonistic example. Of course, most sessions in this period were not all sex, drugs and rock & roll!

Today, the studio availability has almost come full circle. The financial pressures are not so much that the equipment and quantity of personnel are the expensive commodities they once were, or that session availability is limited, but one of sheer business demand. For example, financial advances are far less than they once were, if they exist at all in some extreme cases. As such managers and their artists are beginning to think much more frugally and businesslike about how records can be made. As a result, pre-production has a renewed importance in the production process that can be carried out to maximum effect.

Tony Platt, Engineer, Producer and Music Producers Guild Director says, "[pre-production] has become more important for a lot of artists because it saves money on studio time. I still hear stories however about younger producers going to the studio unprepared and wasting studio time." As such, the lavish and expensive surroundings of a booked studio are usually ditched for a band member's home or a booked rehearsal facility, where concise planning and preparations can take place to maximize the outcome of the more expensive studio time.

As the times are changing, pre-production is one of the most fluid, hidden stages to the outside, non-musical world. It is a process that can and should fit any artist and situation based on the desired vision of the product. Depending on the type of song, album and artist, pre-production can take many different forms that we'll look at in the next section. Therefore, pre-production is the bridge between the composition of the music and the recording session, borrowing skills and tasks traditionally reserved in other stages in the process.

Forms of Pre-production

Originally pre-production took on the form of planning and preparations on the part of the producer. This would involve the booking of the studio, engineers, musicians, perhaps even the catering in addition to anything else that might be required. Attention to detail during rehearsals would be expected, due to the recording sessions often being short, with little or no time for experimentation. Any development of musical arrangement or structure would ideally need to take place in advance in rehearsal. Perhaps the music might have been ready to record straight away.

Today, pre-production has evolved to a more valuable process. Producers have differing views on the importance of pre-production, some preferring to record whole hard drives full of takes from the rehearsal studio to those who simply might arrive at one rehearsal for a quick listen to the band's material before beginning work in the studio. Tony Platt adds, "Sometimes [I record rehearsals] but only very roughly spending time making top quality recordings of rehearsals is a bit pointless. Although sometimes to capture that moment …!"

Traditionally, the pre-production phase can often be considered a vitally important part of the process. This is where so much new songwriting and song development can take place in the comfortable surroundings of the rehearsal studio, coming in at a fraction of the cost of a decent sized studio with an engineer. These are beneficial factors to consider in the current climate where budgets are more restricted. More on this later as we discuss planning a session.

For those artists who do not rely on this traditional structure, pre-production opens up a great deal of opportunity at this stage of the process. Many artists complete much of their work themselves, leaving very little to be achieved in the so-called recording sessions. In Figure 2.1 (bottom), the stages of composition, pre-production and recording blur together and adapt to the project at hand.

Andy Barlow, of the UK duo Lamb, now performing with new project LunaSeeds, reveals that he creates music in such a way. For our music … "the pre-production stage is not really separate from the creative task of songwriting, more it's the oil to the engine of creativity … [now] the only real time there is a big difference is if I am working in Ableton Live. I find this great for writing ideas and getting grooves down. Once this writing stage is finished I will then go ahead and [transfer it and] work on it further … in Pro Tools."

Despite Andy's command of the whole production process, he still prefers to involve specialists at certain stages. Andy's productions often include lavish sonic arrangements that require careful balancing between the occasional orchestral arrangement, real drums, acoustic instruments and electronic samples and synthesizers. All these elements need careful mixing and as such Andy prefers to pass on this responsibility to a mix engineer. "I choose to use a mix engineer for a few reasons: I like not hearing the song hundreds of times [when mixing] as it clouds my vision of it. I get attached to certain sounds because of how long they took me to make, rather than whether they serve the song. I like SSL and Neve Desks (which I don't own). [However] I still like to be very hands on with mixing, and make comments until I am completely happy."

As can be seen in Andy Barlow's example, the creative writing process of the song through to the point at which the material is mixed becomes one big pre-production session, incorporating the writing of the material, the development and intricate programming and production.

AN INTRODUCTION TO CREATIVE PRE-PRODUCTION
An Introduction to Song Development

Developing a song to its full potential can take time and require specific ingredients. To compose a song and walk directly into the studio can extend the length of time it takes for the song to develop to its full potential, needless to say the cost of the project. In extreme cases it can simply reduce the overall quality of the end product. It depends on whether the song is reliant on the recording process and the studio or not. If studio equipment is required to aid the composition, the development and production of a song begins early in the writing process as shown in the lower part of Figure 2.1.

In our "traditional" rock band model, the writing process will perhaps begin by a member of the band at home or on the tour bus and be developed further when brought to the band in rehearsal. Let's dwell on the rehearsal a little.

It is in the rehearsal studio where much of this development and joint ownership of the material can be assumed. Band members are introduced to the new material and will develop new ideas including riffs, syncopations, harmonies, and so on and even new sections. This is where a song can begin to breathe as part of the band "ownership" and not as part of the sole composer. Perhaps this is the reason why there are so many dubious court cases about potential lost earnings from band members' contributions.

Understandably, sole composers within a band often can be somewhat closed off and resistant to ideas that come from others. To exclude ideas from members of the band and outside can close off exciting areas for the song to expand and should be embraced, if not simply given an opportunity to be heard. This will depend on the creative structure of the band and how it develops its material best. Band dynamics and their management are topics outside the remit of this chapter.

At some point throughout the rehearsal, it is worth bringing together the band and the producer. The producer will be able to give a new perspective to the material and offer new ideas to develop the material yet further. For more information on what producers do and whether you need one, see Chapter 1B.

Similarly this can often be considered a "threat" to the composer's ideals and should not necessarily be seen as such. The role of a producer is to ensure that the vision of the artist is successfully transferred to "tape" (sorry hard disk, old sayings die hard) at whatever cost within reason. As such it is important to embrace and welcome the producer's input at this stage; whether you ultimately agree with their ideas are for later discussion.

Some projects will not warrant the cost of a producer. If costs do allow for one, perhaps there will not be enough funds to involve the producer in the rehearsal stage. Such input will naturally take place within the recording studio. However, for some unbiased commentary on the material, some artists call upon the ears of fellow musicians or trusted friends. This can give the band a new perspective or praise for the material as it stands, or some harsh, but true, criticism where necessary.

In addition, it is advisable to capture the rehearsals on anything from a Dictaphone through to a fully fledged ProTools rig. To take away the songs and listen in the cold light of day away from the noise and bustle of the rehearsal can offer excellent objectivity and fresh ideas. We'll discuss this further a little later on.

The Reinvigorated Importance of the Rehearsal Room and Home Studio

Now it's time to consider the way in which the songs will be performed and recorded. This is where we enter into the proper phase of pre-production activity.

Again, this is the time for precise planning and rehearsal. Many bands still believe that the recording studio is the point at which all manner of things can be fixed and made perfect. This possibility of perfection to some extent is true with the onset of advanced tools such as Anteres Autotune, Melodyne and countless others to tame even the most wayward vocalist. Add to this common tools such as Sound Replacer, BFD, and so on for making every kit sound as good as it possibly could do without moving a mic. Wrap this up within the awesome editing capabilities of all Digital Audio Workstations today, the performance can be shaped and crafted in whatever way you might wish. Taking this ability to "correct" a step further, the manufacturers of Melodyne, Celemony Software, have created "Direct Note Access!" (I thought it was an April fools joke at first, but at the time of writing, this feature claims to be able to separate each note in an audio performance for editing in both time and pitch.) This allows incredible advantages to the producer and engineer, but this cannot always make a bad performance good, and in the cases where it can improve a performance, it will not necessarily make the track as a whole better. Nothing can beat a solid performance.

As we've already mentioned, the rehearsal should still remain an important opportunity to improve the structure and arrangement of any song. The emotional architecture within the arrangement using dynamics and harmonic content should be scrutinized and adapted to suit the performing musicians and their foibles. The performances of those musicians playing on the track bringing their own "ingredients" or killer bass lines (as Danny Cope introduced in Chapter 1) to the melting pot should be assessed and developed. As such the rehearsal becomes an important point at which the song can breathe and take shape.

The rehearsal stage should also offer the band the opportunity to tighten up the performance of the material to ensure that the recording session is smooth and ordered. Some bands might wish to play the album material live as a whole set, just to try out the material and to play it solidly in front of an eagerly listening audience.

Rehearsal Recordings

Recording the rehearsals can allow the band to hear how they sound. It is so often the case that the musicians themselves get a different perception of their performance near their amplifiers on stage. Often the band can lose the necessary perspective when constructing and producing a song.

Producers can vary in their approaches if involved in the rehearsals. Some will bring down a small studio setup to capture the musical events. Often, material in the rehearsal studio can have a quality that for some reason or another simply cannot be recreated later in the studio: whether it is an angst ridden vocal take or a particular sound to the guitars augmented by the room; the amp lying around the rehearsal room and player's grubby strings at that time. All this serendipity should be captured! It used to be so often lost to a substandard recording format such as a cassette-based 4 or 8 track machine. The quality of which was normally less than adequate for a final release. As the cost of equipment has reduced, the ability to obtain high-quality digital recordings of rehearsals has become commonplace.

Producers often develop this recorded material and edit it to demonstrate how a different arrangement or parts could improve the music. Development of the material can take many different forms: whether that be messing about with the structure whilst sitting alone with an acoustic guitar to moving parts in a digital audio workstation. This can then be replayed to the band to rule out or take decisions that individuals are discussing. The proof is in the listening.

Either way, the song will have the opportunity to be structured: to be developed; eventually to breathe.

Parameters for Change

The alteration and recomposition of elements such as melody, harmony, and so on are the parameters by which the song development will take its form. The methods and order by which these developments happen cannot be prescribed, and the producer, or other objective listener, will possibly base their opinions on a "gut" feeling. They may hear a different sound or arrangement and suggest changes accordingly. Song development is hinged upon a number of critical criteria from a personal list of things that people look for. Typically the key overarching aspects a producer or reflective artist would wish to alter upon an objective listen back to their material might be:

- Arrangement
- Instrumentation
- Tempo/Time Signature/Groove
- Performance Quality

These four overarching parameters can offer a great deal of scope when altered. Objectivity to music can present a list of things to repair or could spur new ideas as required.

Rehearsing with the band in a loud rehearsal room can be an exhilarating experience and can make material seem solid and workable. However, objective listening at a lower volume whilst using a framework can produce a "list" of elements that will require attention.

Depending on this "list," experiments and alterations to the arrangement, instrumentation (acoustic or electric), groove, etc. can be trialed and might be recorded to listen later. Having a decent recording system within rehearsal allows you to jam around ideas freely and when listening later, elements can be edited together for experimentation.

RECOGNIZING PHASES OF PRE-PRODUCTION

Pre-production has its own internal phases. Phases can be recognized and worked with to monitor progress in a production. Understanding these phases can provide unique points in which to consider the more mundane aspects of preparing for a session (Figure 2.3).

Musical Progress

Phase 1 begins with the songwriting itself, which we assume is a hive of successful creation, and is outside the remit of this chapter. The next stage within Phase 1 is the point at which the individual's material is presented to the rest of the band. The progress may feel as though it is hampered to the composing member, as the band is learning their parts and the arrangement—and beginning to enter into the next stage. This is the initial development of the individual's music by the band as a whole (band-based song development). Parts will be forged, riffs and arrangements created and songs developed as a whole.

Phase 2. Many artists might begin recording at this point and use the studio time to develop their music along a similar path. Phase 2 assumes that the band objectively listens to their material either through some kind of recording of the advanced rehearsals or through traditional demo recording sessions. It is at this stage that comments and constructive criticism will be revealed and might provide an action list. This list, it is assumed, will cause the artist(s) to step back and perceive that their progress has been hampered or set back slightly. Either way, this is an important reality check for the artists as they return to the drawing board a little and potentially carve out a new direction, or alteration to bring the material into line. This will not take a considerable amount of time, but it will ensure that the material is simply ready when the red light goes on. Ensuring that each player knows their parts and can perform in a "second nature" mode, not necessarily having to think about every note, will result in a faster, more productive session.

The band will develop and practice the music until it reaches a plateau of regulated performance quality. In other words, the musicians will repeatedly deliver a consistent and engaging performance of the music they're planning to record either in the rehearsal studio

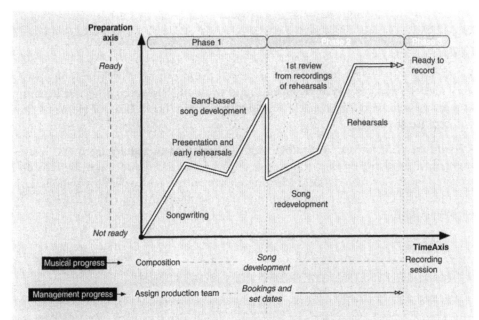

FIGURE 2.3
The perceived progress of a band's material can be mapped as in this fictional example above. Knowledge of the band's musical progress can perhaps also act as a trigger for management activities within pre-production such as booking musicians, studios, mastering, and so on.

or a live performance. It is at this point that the band is ready to record and enter into Phase 3.

Management Progress

If a band's progress can be tracked then it is possible to predict when the material is in its second phase. It is at this point that the music becomes galvanized into something new using the information provided on the "list," or reverts back to the better original format in Phase 1. At the start of Phase 2, the material is brimming with new ideas and may be forged into its final state. It is during this time that studios, session musicians and other bookings could be made with some form of certainty as the format begins to settle down.

This certainty provides the artist, producer and production team an opportunity to plan Phase 3, the recording session, properly. It is at this point that musicians will be confident of the elements they will wish to include in the songs, and therefore the equipment, personnel and space needed to produce the music in question.

The Ability to Listen Objectively

To make an accurate assessment of music, it is important to develop certain skills in listening. To be able to do so in an objective manner is imperative before developing songs or choosing to throw them out.

©2009 Russ Hepworth-Sawyer

We all listen to music for enjoyment, but it is extremely difficult to do so in an unbiased and detached fashion. As Danny Cope outlined in Chapter 1, a song should have "influence" and naturally draw the listener in. Thus to measure this "influence," the art of listening objectively should be developed, as it is difficult to remain focused on how the music will appear to the listener when taking part in its development over time.

To have the ability to switch between detached (passive) and attracted (active) listening to music is key here and is a skill that is paramount to be in a position to shape music accordingly. This should be something we should be able to do for a particular audience whether we like the material or not. Objective listening can take time to develop. Audio professionals such as producers and engineers spend a professional lifetime listening with an objective ear.

In a similar fashion to that of classical composers who can imagine or emulate the sound of the orchestra in their heads, many audio professionals can also hear music altered and edited, engineered and polished before any equipment has been operated. This ability can be developed and is something that first begins with concerted analytical listening and then can be switched off into detachment as required. A detachment that means the listener analyzes "sound" on occasion as opposed to music. The exact process by which this happens is outside the remit of this book although a brief example is provided in "Listening Frameworks for Analysis".

Working in music and audio brings along an allied objective ability. Strange, but useful, is the ability to listen to the same song, verse or guitar part over and over again without tiring of it, whether we like the piece or not. Being able to detach yourself from this may seem a little like masochism to the casual listener, but is imperative to work as we do whilst attempting to remain objective.

Later, a model for planning is outlined for a production and listening skills such as these enable us to identify whether or not the material is on the same road as we had planned. The framework discussed later may not be detailed enough to "guide" the listener to immediate objective qualities to make or break a track. Nevertheless, an improved listening ability should ensure that problems in a composition or its arrangement could be identified and rectified.

Whittling Down the Material

Choosing which tracks are to be recorded can happen at various points. At the end of Phase 1 to the start of Phase 2, the band may listen on reflection to some of their material, which might be a joy to perform, but as a stand-alone song, will not engage with the fan base. More recordings can be made of the songs throughout Phase 2 to assess the worth of pursuing with the material. In most cases these songs will move forward to the session and may be used as B-Sides if not selected for the album.

So many bands work on many more compositions than make the final track listing on a CD. In many cases it is not uncommon to rehearse and record many more songs for a CD than the typical 12–15 tracks. Working with a producer, or taking on the duty as a band, the material will begin to get sieved and perhaps at the same time the singles will begin to present themselves, perhaps alongside a track order, as the material gains a persona of its own.

This whittling down of material is an iterative process and takes time. Thus adequate planning should be given to returning to the drawing board after the first phase of the pre-production process. When this point is reached will be dependent on the band and the composition team within the project.

PRE-PRODUCTION IS TO PLAN!

So often the word "planning" seems to grate against most people's view of what record production is. Planning and project management are mandatory for businesses and many walks of life. It is a serious and expected skill. One would not take on a business venture with a large budget and various personnel involved without considering the legalities, financial position and meticulous planning. So why would a music production venture be vastly different, provided enough leeway is given to the creative process? Planning can take many forms and is explored in this section with specific reference to music production.

Planning the project from start to finish might seem a very dry and restrictive thing to do, but a necessity to some extent. It is common to read about sessions with the Beatles in Abbey Road seemingly lasting for weeks on end. Clearly an exaggeration, but most of us will not get similar opportunities in today's commercial studio and as such there needs to be a determined end point—a delivery date. The world has changed a little and things move at a considerably faster pace nowadays.

So much is based on the finances of a project, as the labels have less money to throw about. Worse still, the artist might see themselves bearing the costs more and more as new models of distributing music become apparent and record deals are slowly adapting. As such, planning takes on a new, more poignant importance to ensure that the project comes in on budget.

All projects have driving parameters for completion, most usually a deadline and a budget. These two parameters have an overriding effect on how the project should and can succeed. These could be seen as restrictions, but can also act as serendipitous opportunities in the sense that the design of the plan can lead to creative and new thinking that could give your project an edge over the competition.

Some projects will take next to no planning whatsoever as the artist may have their own studio facility and like to play all the instruments. In this situation, it is simply a matter of planning when to get in the studio together with the producer to crack on. There will always need to be a plan as the producer is likely to have prior commitments for the next album they may be working on. In the same vein, the artist may have touring commitments too that need to be factored in.

However, many projects will simply take a little more time to plan and organize based upon the complexity of the music in question. Projects that need to include session musicians need to be factored in as well as any arrangers and orchestras you may wish working on the tracks.

In either of these scenarios and many more, a structured starting point might be employed. For this purpose, V.I.S.I.O.N. has been produced. This can prompt the team into considering a number of aspects about the project to ultimately agree on the vision for the product. This trigger list is the halfway house between the creation of the project and the project management to come. Please refer to the V.I.S.I.O.N. box explained below.

A V.I.S.I.O.N.

Going straight into a studio can be an excellent way of getting started and developing music. It is fair to say that most would all love the opportunity, but as we've outlined above, this is not always possible depending on the size of the project. As such, some direction is useful to ensure that an end goal is achieved. Succumbing to your business mind again will ensure you plan accordingly.

Any such plan must follow a path that has been agreed and mutually understood by all concerned, otherwise how would we know we had got there or which road to travel upon? To plan toward an end product, a vision needs to be considered, or at least the anticipated outcome needs to be visualized.

To assist this, a simple checklist called the V.I.S.I.O.N. is provided that addresses many of the aspects that could be conceptualized before beginning a project. Securing the *vision* of a project will empower everyone to talk from the same hymn sheet. It is important that this vision is easily disseminated to others and understood with as little misinterpretation as possible.

On page 34, Figure 2.4 details the V.I.S.I.O.N. trigger list. Below we dissect each area it covers as you plan a music production.

Visualize: Visualizing the end product is not necessarily just a dream of how the material may sound, but should also be a target position in its relative marketplace. An understanding of the current music scene and how this will slot in will be of use when considering the material and what direction it should take. Consider that of any genre that is constantly evolving such as hip hop or dance music. These changes vary frequently and visualizing the music's position within that marketplace will not only give pointers to where it should or could fit, but also provide the suitable inspiration to differ and innovate.

Innovate: With each record that is released, innovation is often at the core and musical genres develop almost as fast as computer processing power growth as stated by Moore's Law. Whilst innovation might not be the sole aim of the game for the artist, it will, by its sheer virtue, be something new. Whether this is a radical departure for the band and thus alienating its audience a little, or whether it plays it too safe to keep hold of the fan base, it will still be new material. The million-dollar question is how a band can generate enough interest album after album. New music needs to inspire and draw listeners in as described by Danny Cope's

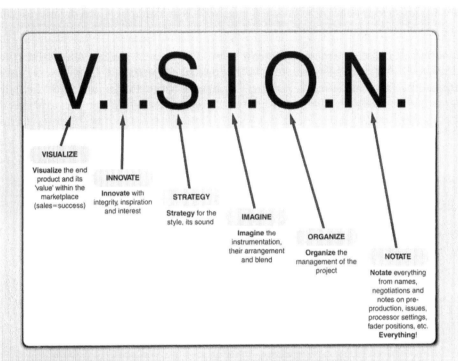

FIGURE 2.4
The V.I.S.I.O.N. is an acronym we use to remind us (or "we can use to prompt considerations of" ...) of the many elements that might need to be planned in pre-production.

triangle of influence. Creating this interest can be drawn from inspiration, but can be delivered when working with a specific producer. Thus, the producer can have a profound influence on the music.

Being in touch and having an ear to the ground musically will influence a different and fresh perspective to the material. An awareness of the genre and scene will also provide a much valued success gauge. Thus, knowing if this tune were to do well in the charts and clubs or not, or whether more work would be needed to get it to the same standard or exceed it. As such it is not simply how it sounds but its value within its genre or marketplace. This is connected with the style and sound of the work, which we cover a little later.

That said, it is important to consider the origin of the material and treat it with integrity and not follow a personal route. The fellow band members or producer should take into account the wishes of the composer and try to bring the material to life without altering it away too far from its origins without agreement. Sensitive musicians will present some of the hardest challenges within the studio and such integrity to the music and the cause is an important pursuit for all.

Strategy: Based on the value judgments and assessment of the genre and scene that have been made above, it is important to produce a strategy for the style and sound of the recording. This strategy will be an outline of the kind of sound you're wishing to achieve and will point you in the right direction in the organization and planning of your project.

Style in this example is meant not only to be about the sound, but the way in which the band looks and acts. The charts are usually full of what is known as "manufactured" artists that have been styled already. This is an important facet and a clear selling point, and in some cases it will begin new trends in society for dress sense. This "style" may not simply be about the clothes that are worn, but the way in which the artists carry themselves.

Imagine: Imagining the end product is one thing, but noting the instrumentation that goes up to make the vision is crucial in the planning. Much of the instrumentation may be dictated by the songwriting itself and be somewhat dependent on the parts, such as a riff-based song. To picture the blend of the instrumentation together can lead to new sounds and ideas which can lead to new production flavors.

At this point, it is also worth considering arrangements of music and their position in the track listing as songs. During production this will invariably change, but it will assist the planning and the "list" of things the band needs to achieve before recording and beyond. For example, a concept album will require to work in a certain order such as *The Wall* by Pink Floyd.

Organization: Pre-production will involve a healthy amount of this I'm afraid. It is during this stage of the production process that the vision is determined, or crafted, rehearsals are about to start or are in swing, and the planning needs to happen in earnest based upon the vision for the music. To begin to consider the production project as a whole and how it will come to market is important at this point. Choosing the production team, the studio, etc. is all part of the planning process.

Notate: Many bands and indeed studio professionals fail to notate as much as they should. Bands may forget older songs that have fallen by the wayside. Without records, these compositions could be lost. The same works for the logging of settings, agreements, conversations, ideas and notes. Disciplined logging is rather useful in planning large projects. It's a business after all.

Engineers and producers should take notes about all-important effects processor and compressor settings for a big mix session they're trying to recall. Notation such as this is far easier these days with recall on pretty much all-modern equipment, but still, there will be the odd coveted Urei 1176 or Pultec EQ that will need to be marked off on a template somewhere. For more information about studio notation, see Chapter 4 by David Miles Huber and Robert Runstein.

Notating the event through video or photographs is also something that can be used for the band's or production team's website, or career materials for the future. If they never make it into the public domain, they will be something to view with interest in the years afterwards.

By using the prompts provided by the V.I.S.I.O.N., answers to prominent questions such as these below should be forthcoming. Alternatively, these questions should at least prompt the team to consider these areas further until a solution is reached.

- *Visualize*
    - Is there an agreed visualization or feel for the end product? If not, how can this be reached?
    - Is there a concept behind the album, a story or just a collection of songs?
    - How will this product fit within the genre now? In 6 months' time?

- *Innovate*
  - □ Are there any benefits to adding some new innovations to this record?
  - □ What could these innovations be?
  - □ Will they alienate current fans or will they act as an influence for new listeners?

- *Strategy*
  - □ What strategy can be put in place to create the visualized sound that is wished for?
  - □ What reference points can be drawn to achieve this end product: *sentimental music with comedic lyrics about Mickey Mouse*?

- *Imagine …*
  - □ … the instrumentation and consider this as part of the overall visualization: *mainly a drum'n'bass beat with operatic vocals? Perhaps heavy metal guitars, bass and vocals and a tambourine instead of a kit*?
  - □ … the arrangement or again style of arrangement?
  - □ … the sequencing (the order of the material on the album)?

- *Organization*
  - □ Can you begin to consider the planning of the project?
  - □ Who are the production team to be?
  - □ What rooms and studios can you use/do you want to use?
  - □ Will you need external musicians and equipment?
  - □ What planning can you make at this stage about the initial release?
  - □ Is there a specific date you'd want to release this: *Christmas? FA Cup? World Cup? Olympics? April Fools Day*?
  - □ How do you want the product to be presented based upon the answers to the concept and visualization above?
  - □ Artwork ideas?
  - □ Packaging?
  - □ Management of the digital downloads?

- *Notate*
  - □ Can you confidently remember everything? If not, create a notation scheme you trust so that planning and thoughts can be captured.
  - □ Prepare notes on all aspects of the production that pertain to you.
  - □ Can you prepare notes about the material for the production team?
  - □ What else can you notate and provide for the rest of the project team about the music and the project as a whole?

PROJECT MANAGEMENT

Taking on board the answers to questions like the ones above about the project, one should turn to the planning of the production. So what planning should be considered? Figure 2.5 considers a range of elements.

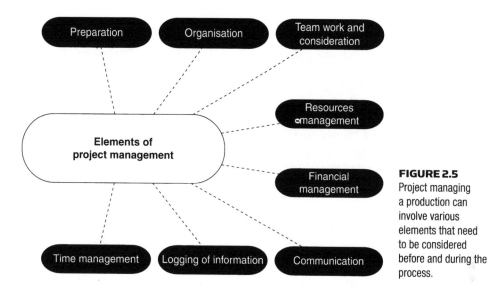

FIGURE 2.5
Project managing a production can involve various elements that need to be considered before and during the process.

Preparation and Organization

Before the recording sessions where the big money might be being spent hiring a studio, it is imperative that the artist(s) and the production team are all speaking from the same hymn sheet—the same vision. Sensible preparation with meetings between all personnel concerned can save valuable studio time at a later date. David Huber & Robert Runstein speak more about preparation for the sessions in Chapter 1B.

The remaining preparation is to engage with the rest of the project and plan accordingly....

Being organized has never been hip and trendy, but is frankly essential in a professional environment. It ensures that goals are reached and the success of the project as a whole. This organization will rub off on the team and the project as other areas of the project management are dealt with.

Team Work and Consideration

The team is paramount to the success of any project. The team surrounding the project should have an internal dynamic that will drive the music forward and the production as a whole. Understanding team dynamics and the way in which the project can be best realized within those parameters is an important facet. Most producers will be masters at ascertaining the skills and personal dynamics between the band members. They will have many tricks up their sleeves to be able to work with a variety of personalities in a range of situations. They will consider the process and be continually sensitive to the needs and feelings of the band. It is a joint venture after all.

Resources Management and Financial Management

As with any project money must be spent on resources, whether that be equipment, facilities or personnel. These two are clearly intertwined, and thus the management of the budget is imperative to the outcome of the project as it will restrict or allow the flexibility possible. The manager and producer will preside over these issues, but in many cases it might be absolutely necessary for the musicians to manage the money, especially if self-funding and self-managing. In either eventuality planning will aid the ultimate outcome of the project at the best price, thus ensuring more profit for the team. More detail on this matter is outside the remit of this book.

Communication and Logging of Information

As we've already mentioned, communication with the team and all concerned with the production is absolutely vital in ensuring the project runs smoothly. Logging things such as conversations, notes, settings and other elements is a wise undertaking as it allows accurate recall of information and projects. There are many ways this can be achieved which, again, is outside the remit of this book, although some productivity methodologies have become popular in business in the past few years, such as David Allen's *Getting Things Done* (GTD®) and others found in the bibliography list at the end of this chapter. These systems are very flexible and can be excellent pointers in improving your productivity in whatever field you are in. It is well worth taking time to explore these.

Contingency and Time Management

The V.I.S.I.O.N. and the questions coming from it should provide a guide as to what could be considered, and in what detail, before a production is undertaken. It is important to be aware that this seemingly rigid and business-like mentality needs enough flexibility to permit fluid creativity and expression throughout a project. In order to accommodate such creative expression, one should consider some contingency when planning.

Of course there would be problems with committing everyone to anything too rigid as ideas will flourish and avenues will present themselves in the studio that were never there before. These ideas take time and as such could potentially put your time plan in jeopardy. If the schedule slips, that string quartet that was booked might knock on the studio door as you're still tracking the rhythm section due to the song developing in another area. Therefore it is crucial to consider placing in contingency for such things in the time plan.

Whatever the budget, studios need to be sourced and booked along with the production team. Session musicians may need to be booked and hired too. The time frame needs to slot in with the key personnel you wish to involve in the project and in such a way that the costs are kept under control. So there is an immediate balancing act between what should be done, what might be done and the strict milestones that will need to be met.

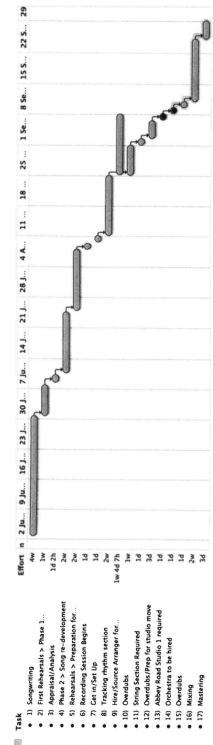

Task	Effort
1) Songwriting	4w
2) First Rehearsals > Phase 1...	1w
3) Appraisal/Analysis	1d 2h
4) Phase 2 > Song re-development	2w
5) Rehearsals > Preparation for...	2w
6) Recording Session Begins	1d
7) Get in/Set Up	1d
8) Tracking rhythm section	2w
9) Hire/Source Arranger for...	1w 4d 7h
10) Overdubs	1w
11) String Section Required	1d
12) Overdubs/Prep for studio move	3d
13) Abbey Road Studio 1 required	1d
14) Orchestra to be hired	1d
15) Overdubs	1d
16) Mixing	2w
17) Mastering	3d

FIGURE 2.6

Gantt Charts are used in project management to organize fairly large undertakings and any production project could be mapped like this. The benefit of undertaking such a formal process can illustrate the flash points in the project, as well as quickly and easily demonstrate the plan to the rest of the team.

In business, project managers use elaborate software to organize personnel and activity against time. This can quickly demonstrate what resources are needed at any such time in the project. Most of these systems produce charts called Gantt Charts that allow you to quickly track the activity of the project. Many producers have much more to worry about than creating such elaborate charts, but may of course keep a mental note (or at least a diary entry) of the milestones along the way such as "Abbey Road Studio 1 for orchestral recording." Such markers will therefore need to be signposts, ensuring that the music and parts are written and ready before this expensive session (Figure 2.6).

However the project is planned, whether software, Filofax or iPhone, it does not matter as long as everyone is on top of the elements and are in a position to ensure it runs smoothly. However, a Gantt Chart does enable a quick and visual representation of the project and how each element slots into the process.

It is important to ensure that everyone involved is kept up to date about slippages or any issues with the time plan as it stands. Communication is such a large part of the process, and ensuring that the project team are kept informed is clearly paramount.

BEFORE THE PRODUCTION

In this chapter, we've discussed pre-production in an academic manner. Pre-production can happen in so many different ways. Every eventuality could not be written in one short chapter such as this. The information here is transferable to many differing genres and styles of music. Planning is always an essential element to the successful execution of a project.

Much of this planning will usually be the remit of the production team, but invariably it will be important for the musician to play a large part in this, after all it is the artist's money that is being spent, even if it is from the label's advance.

FURTHER READING

Allen, D. (2001). *Getting Things Done*. Piatkus, London.
Crich, T. (2005). *Recording Tips for Engineers*. Focal Press, Boston.
Forster, M. (2006). *Do It Tomorrow and Other Secrets of Time Management*. Hodder & Stoughton, London.
Huber, D. & Runstein, R. (2005). *Modern Recording Techniques*. Focal Press, Oxford.
Lewisohn, M. (1989). *The Complete Beatles Recording Sessions*. Hamlyn/EMI, London.
Moylan, W. (2007). *Understanding and Crafting the Mix, the Art of Recording*. Focal Press, Oxford.
Stone, C. (2000). *Audio Recording for Profit: The Sound of Money*. Focal Press, Boston.

Chapter 2

B. Listening Frameworks for Analysis

Russ Hepworth-Sawyer

In This Chapter

INTRODUCTION

Developing a Listening Framework for Analysis is a way in which analytical listening in music production can be taught. This short chapter is inspired by the excellent work of William Moylan.

EXAMPLE: LISTENING FRAMEWORK

Frameworks for analysis are the starting point to dissect how a recording has been put together objectively and how it's result actually sounds. Figure 2.7 is an example of what we might consider the bare essential elements and a structure by which listening can be analyzed. This framework will need to develop and mature depending on its use and function.

Listening with a critical ear is a personal pursuit and will vary widely from person to person. In the framework example below, we have introduced the idea of using listening passes (passes refer to amount of times the tape passes the playback head or the amount of times we listen to the piece). In this example we listen 3 times through and break these main areas down yet further to assess them and comment upon them. It should be mentioned that more passes are often required.

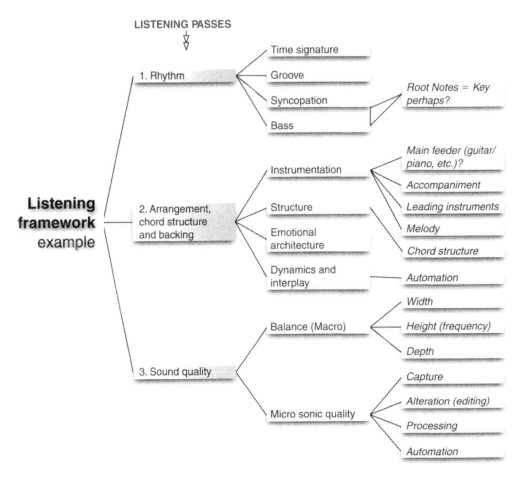

FIGURE 2.7
A listening framework such as this basic example can provide a checklist against which overall sonic and production qualities can be measured or checked.

This branch diagram could be expanded to offer exacting questions from each of the prompts here if required. With more experience one or two passes should suffice for the simple list shown above.

Pass 1

In this example, pass 1 is entitled Rhythm. The first thing that can be assessed is the form of the drums or rhythmical elements. Analyzing this part of the recording can reveal the time signature and offers some insight into the groove of the piece. Taking this one step further, linking the bass of the song within this pass, we can learn much about how the song is constructed, whether there is syncopation, and what the bass is playing. This pass could of course be expanded to the types of rhythms and how they interact with other instrumentation.

Pass 2

Pass 2 turns attention to the more musical aspects of the recording. In this pass we're concerned with the arrangement, chord structures and backing—essentially the way in which the music is constructed. This includes an analysis of the instrumentation. It is here that an identification of the main feeder can be made. The main feeder in a song is the instrument that drives the song. Not a vocal hook necessarily, but an element that makes the song both memorable and often the main driver in the backing. It may be the catchy riff or the solid pounding of piano chords. We look here also at the accompaniment—what makes the whole composition tick behind the scenes sonically. Finally the melody can be assessed.

The arrangement section is broken down into three elements: structure, emotional architecture, and dynamics and interplay.

The structure of the piece, if it is your own composition, is already known, but as new tracks are introduced and these skills develop, structure will be an informative place to gather plenty of information of the construction of the song. It is at this stage we can identify if there are things that can be changed and manipulated.

We next assess the music's emotional architecture. Emotional architecture is the way in which the music builds and drops. This is not an assessment of dynamics (discussed later), but an assessment of how the music affects the listener through its intensity. This is something that can be drawn graphically as shown in Figure 2.8.

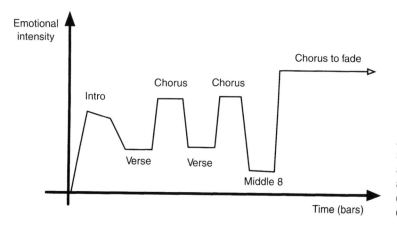

FIGURE 2.8
Emotional Architecture Graph showing the intensity and power of a track as it delivers its choruses and a large ensemble at the end.

In this graph, the emotional intensity or power of the music can be shown and tracked. Assessment of many popular tracks can often show themes or perhaps new ideas of how to achieve this variety in emotional intensity.

Finally we would quickly assess the dynamics of the instrument and how this interplays with the track as a whole. So, for example, the dynamics of a given instrument and how it might be integrated into the mix using automation or other mixing techniques are covered later in this book.

Pass 3

On the last pass, we assess the sound quality of the music and dissect its construction. This is split into two areas and here we introduce the terms "Macro" and "Micro." In this case we refer to Macro, meaning the balance of the instruments together, and Micro is the focus upon individual elements within the mix and their sonic quality.

Macro listening to the balance delivers three sub-sections which are fairly broad: Width, Height and Depth. The first refers to stereo width and the use of panning. We would wish to make an assessment of how the stereo field has been made and constructed. Height, in this instance refers to the use of the frequency range or spectrum. Some artists make full use of all the frequency ranges with deep, but focused bass and excellent detail at the top end. Whilst other pieces choose to limit the overall bandwidth they occupy. It must be noted here that Height can also refer to the vertical presentation of a sound. Finally is Depth, which is an assessment of the presence of any given instrument or the mix as a whole. Is the mix deep, presenting instruments that appear to come from behind others, or are all instruments brought up to the front in the mix?

Micro listening is an assessment of the internal elements that make up the sounds in our mix. How instruments have been captured and whether there are issues inherent in their sound can be gauged. Listening in a focused manner using a good reproduction system can also reveal edits in material that were not supposed to be heard, drums that were replaced with samples and vocals that were automatically tuned. Estimates of the types of processing that were used in achieving the mix can be speculated.

Automation now plays a huge part in mixing and presenting the micro elements. Can an assessment of the automation moves be assessed upon very close inspection? How does this shape the presentation of the music?

FRAMEWORK ANALYSIS

Once an overview of a framework is decided upon, we advise that comments are scribbled down in response to each of the areas in the framework. A personalized pro forma could be created. An example has been given below based on the basic framework shown in Figure 2.9. Any pro forma document should be well designed in such a fashion so it takes little time to become accustomed to, allowing reliable information to be recorded about the song in question. How this is notated will be a personal method based on whatever can be notated in the boxes.

In time, this skill should not require a lengthy written document or framework pro-forma as suggested here, but some to jot down simple notes of pertinent points observed. This framework acts as a checklist that one day should become second nature.

In the version below, a student's rough analysis is shown using a 1978 Genesis Track called "Follow You, Follow Me," it can be seen that this analyst has chosen to write comments that might be pertinent to the track's qualities. For example, these have

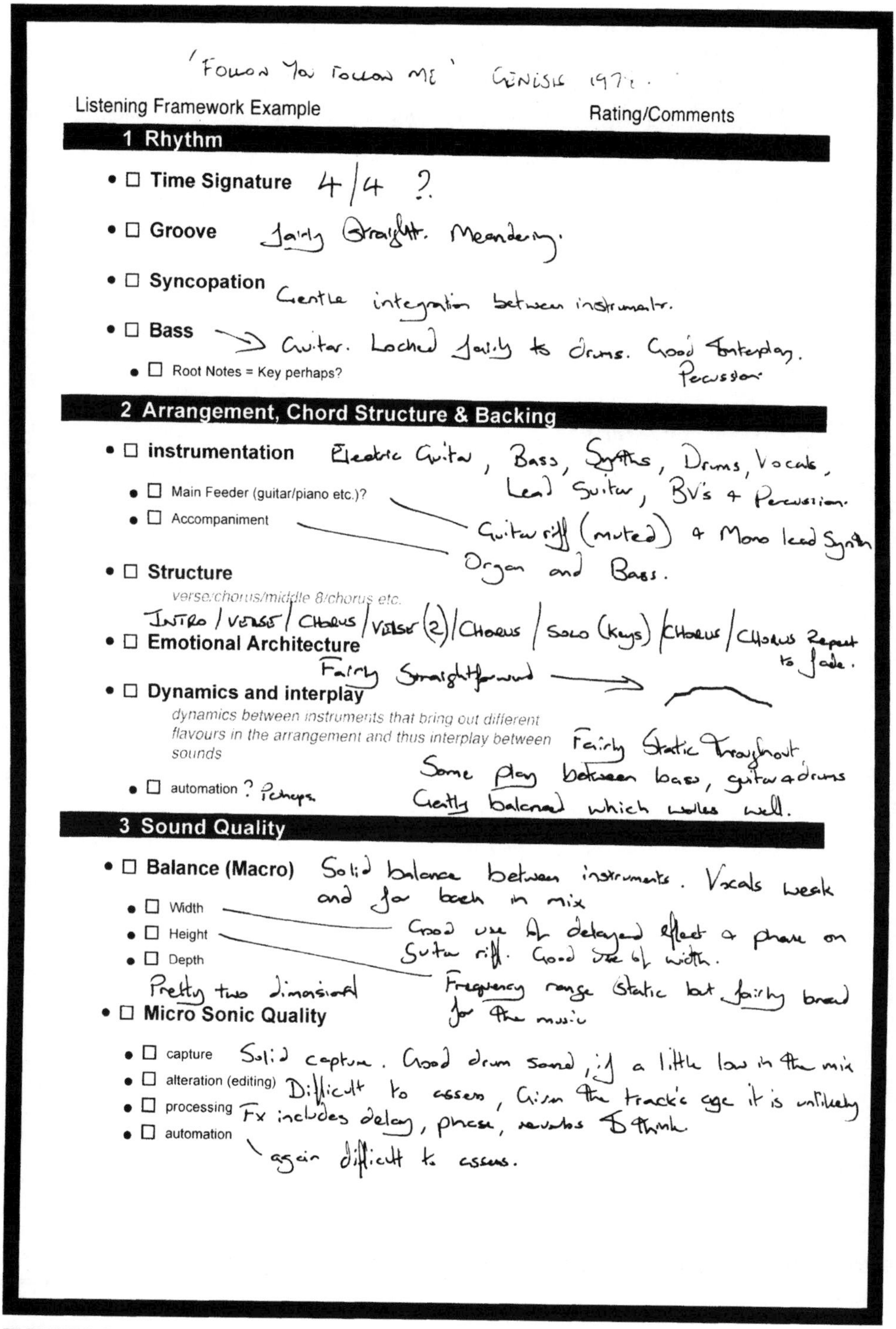

'FOLLOW YOU FOLLOW ME'   GENESIS 197?.

Listening Framework Example   Rating/Comments

1 Rhythm

- ☐ **Time Signature** 4/4 ?
- ☐ **Groove** Fairly Straight. Meandering.
- ☐ **Syncopation** Gentle integration between instruments.
- ☐ **Bass** → Guitar. Locked fairly to drums. Good Interplay.
    - ☐ Root Notes = Key perhaps?   Percussion

2 Arrangement, Chord Structure & Backing

- ☐ **instrumentation** Electric Guitar, Bass, Synths, Drums, Vocals, Lead Guitar, BV's + Percussion.
    - ☐ Main Feeder (guitar/piano etc.)?
    - ☐ Accompaniment Guitar riff (muted) + Mono lead Synth Organ and Bass.
- ☐ **Structure**
    verse/chorus/middle 8/chorus etc.
    Intro / Verse / Chorus / Verse (2) / Chorus / Solo (keys) / Chorus / Chorus Repeat to fade.
- ☐ **Emotional Architecture** Fairly Straightforward →
- ☐ **Dynamics and interplay**
    dynamics between instruments that bring out different
    flavours in the arrangement and thus interplay between   Fairly Static Throughout.
    sounds   Some play between bass, guitar & drums
    - ☐ automation ? Perhaps   Gently balanced which works well.

3 Sound Quality

- ☐ **Balance (Macro)** Solid balance between instruments. Vocals weak and far back in mix
    - ☐ Width
    - ☐ Height   Good use of delayed effect & phase on
    - ☐ Depth   Guitar riff. Good use of width.
    Pretty two dimensional   Frequency range static but fairly broad for the music
- ☐ **Micro Sonic Quality**

    - ☐ capture Solid capture. Good drum sound, if a little low in the mix
    - ☐ alteration (editing) Difficult to assess, Given the track's age it is unlikely
    - ☐ processing Fx includes delay, phase, reverbs I think
    - ☐ automation
        ↘ again difficult to assess.

FIGURE 2.9
A student's analysis using a Listening Framework Marksheet.

become a list of things to listen out for in the next pass such as the key is in some dispute. These have been adorned with question marks and can be reference points for later investigation on additional passes. Other listeners might simply wish to "rate" the sound and use a star rating, although this would not give exacting points for discussion or further analysis.

As discussed, this example, or method, is far from exhaustive and should be expanded and personalized where possible to suit the genre and purpose of the analysis. As such we would recommend to develop this in close conjunction with Chapter 13 by Bruce and Jenny Bartlett.

To develop this concept much further and to a more thorough level, see Moylan (2007).

FURTHER READING

Bartlett, B. & Bartlett, J. (2005). *Practical Recording Techniques*. Focal Press, Oxford.
Moylan, W. (2007). *The Art of Recording: Understanding and Crafting the Mix*. Focal Press, Oxford.

Chapter 3

Capturing, Shaping, and Creating the Performance: The Engineer's Role in the Process Ahead

William Moylan

In This Chapter

- Introduction
  - □ Re-cap on Pre-production
- Recording and Tracking Sessions: Shifting of Focus and Perspectives
  - □ Analytical Listening Concerns
  - □ Critical Listening Concerns
  - □ Anticipating Mixdown During Tracking
- Signal Processing: Shifting of Perspective to Reshape Sounds and Music
  - □ Uses of Processing
  - □ Listening and Processing
- The Mix: Composing and Performing the Recording
  - □ Composing the Mix: Presenting, Shaping, and Enhancing Musical Materials
  - □ Crafting Musical Balance
  - □ Pitch and Sound Quality Concerns in Combining Sound Sources
  - □ Creating a Performance Space for the Music and the Recording
  - □ Performing the Mix
- Summary
- Further Reading

In this chapter, William Moylan discusses the elements
that the production team are looking out for and
how they listen. At this juncture, we look at some
pertinent issues of sound in the recorded form, which
should be addressed or considered in all sessions by
many members of the production team and the artist(s).

INTRODUCTION

With objectives clearly defined and a vision for the sound and content of the project, the recording can now take place.

The engineer will capture the performance through recording or tracking the musical parts and instrumentation, and will shape sound qualities and relationships of the performance through equipment selection and signal processing. The mixing process later will continue the shaping of the performance and will result in the creation of the final mix.

The actual act of recording sound begins here. The preliminary stages that were discussed in the previous chapters will define the dimensions of the creative project and of the final recording. In addition to these, certain other preparations must take place before a successful tracking or recording session can begin.

Recap on Pre-production

Whenever possible, the musicians should be prepared for their performance at recording and tracking sessions. Unfortunately, it is possible some musicians will get their music or performance instructions when they arrive for a recording session. This limits their potential to contribute to the musicality of the project and can minimize the quality of their performance.

Ideally, music should be given to performers well in advance of a session. Solo performers should have the opportunity to learn their parts before they join the other musicians, and all musicians expected to perform simultaneously as a group should be well rehearsed together (as an ensemble). As much as possible, musicians should be given the opportunity to arrive at the session knowing what will be expected of them, know their parts, and be ready to perform. It is often best when the music to be recorded has been performed several times in concert—in front of an audience. In the reality of music recording, this ideal is not always feasible.

Some final instrumentation decisions and many of the supportive aspects of the music are, however, often decided upon during the preliminary stages of the recording process (or during the production process itself). The engineer will learn to anticipate that some musical decisions will be made during tracking. They will become aware that dimensions of creative projects can tend to shift (sometimes markedly) as works evolve—especially between types of clients and types of music. The engineer and production team must keep as many options open as possible, to keep creative artists from being limited in exploring their musical ideas. They must not find themselves in the position of not being able to execute a brilliant musical/ production idea (without hours of undoing or redoing) because of a recording decision made a few minutes earlier. Allowing for flexibility may be as simple as leaving open the option of adding more instruments or musical ideas to the piece, by leaving open tracks on the multitrack tape or in console layout, cue mix changes and signal routing, or in mixdown planning.

The recording studio is the musical instrument of the production team. The recording process is the musical performance of the engineer. In order to use recording for artistic expression, the engineer must be in complete control of the devices in the studio and must understand their potentials in capturing and crafting the artistic elements of sound.

RECORDING AND TRACKING SESSIONS: SHIFTING OF FOCUS AND PERSPECTIVES

Recordings are made with performers recorded individually or in groups, or with all parts of the music performed at once. "Recording sessions" as used here are recordings of all the parts being played at once; the entire musical texture is recorded as a single sound.

Tracking is the recording of the individual instruments or voices (sound sources) or small groups of instruments or voices, into a multitrack format (digital audio workstation, analog multitrack recorder, etc.). This is done in such a way that the sounds can be mixed, processed, edited, or otherwise altered at some future time and without altering other sound sources. For this isolated control of the sound source to be possible, it is imperative that the sound sources are recorded with minimal information from other sound sources and at the highest, safe loudness level the system will allow.

Recording and tracking sessions require the engineer to continually shift focus and perspective, while listening to both the live musicians and the recorded sound. Further, they will move between analytical and critical listening processes. The musical qualities of the performance will be constantly evaluated at the same time as the perceived qualities of the captured sound. The production team is responsible for making certain both aspects are of the highest quality and exist accurately and without distortion state in the storage medium.

Focus will have the listener's attention moving freely but deliberately between all of the artistic elements of sound and all of the perceived parameters of sound. The engineer will need to shift focus between artistic elements (perhaps shifting attention between dynamic levels, pitch information, or spatial cues), then immediately shift to a perspective that evaluates program dynamic contour (of the overall sound). They are required to continually scan the sound materials to determine the appropriateness of the sound (and its artistic elements) to the creative objectives of the project, and to determine the technical quality of the perceived characteristics of the sound.

Within the recording and tracking sessions, the engineer is concerned about the aesthetics of the sound quality that is going to "tape," and they are concerned about the technical quality of the signal. Depending on their function in the particular project, they may not be in the position to make decisions related to performance quality. They should nonetheless be aware of the performance and be ready to provide their evaluations when asked or to anticipate the next activity (repeat a take or move on to the next section, as examples). It is imperative that the engineer always be aware of the technical quality of the recording. It is their responsibility for high-quality sound to be recorded.

Analytical Listening Concerns

There are many performance quality aspects of music that will engage the production team in analytical listening at this stage. Most of these aspects are related to:

- intonation,
- control of dynamics,
- accuracy of rhythm,
- tempo,
- expression and intensity,
- performance technique.

The pitch reference, time judgment, and musical memory skills gained in a studio-based career will now be invaluable. Staying in tune, at a specific reference (such as A440) is necessary for "Take 1" to be useable against, say, "Take 617." Tempo also must not change unintentionally and dynamics must remain consistent throughout the session(s). Dynamics can be altered by performance intensity, and these changes in expression should be under control and factored into dynamic level considerations. How a musical idea is played is often just as important as the idea itself; the expressiveness of the performance and its suitability to the project are important aspects of great recordings (such as the ones we seek to make). While these areas may not always be under the purview of the engineer, flawed material needs to be identified at all times. Just how and when issues get pointed out to the artists, producer, or others varies with the project, roles, and personalities involved.

The performance technique of musicians is critical in recording. The performer's sound cannot be covered up by the other players or assisted by the acoustics of a performance environment, and will be apparent in the final sound. The ways performers produce sound or the instruments themselves can create the desired musical impact, or it may not. The person responsible for the project will need to make evaluations and to offer necessary alternatives. The producer and engineer should know any performance technique problems and the natural acoustic sound properties of all of the instruments (whilst well outside the scope of this writing).

At this stage the engineer should be getting a clear idea of the intended overall qualities of the final recording, and how each track will contribute to those overall qualities. Tracking will capture or shape sound significantly, and will either allow the engineer to achieve the desired sound or it will fall short. Careful attention must be given to the sound quality of the sound source for timbral detail, performance intensity, dynamic contour, spatial concerns, sound quality (for pitch density), and more. Depending upon what source is being recorded at the time and the final intended sound quality, some artistic elements might be more important than others; still all contribute important information and must be considered.

Critical Listening Concerns

The technical quality of the recording will be reflected in the integrity of the recording's signals. The recording process must not be allowed to alter the perceived parameters of sound, unless there is a particular artistic purpose or technical function for the alterations. The perceived pitch, amplitude, time elements, timbre, and spatial qualities of the recorded sound sources should be accurately captured in the recording and should be the only sound present in the reproduced recording. Any extra sounds are noise and any unwanted alterations to the sound are distortions. The engineer must know the recording process and devices instinctively well for this to be accomplished fluently. The goal is for impeccable technical sound quality to be achieved effortlessly, so that attention can be mostly given to the artistic qualities of the sound and the performance.

Accurate recording and reproduction of the audio signal must be consistently present for professional quality music recordings. Skills in evaluating sound quality, time information, and dynamic contours at all levels of perspective are important here. Sound will be evaluated for flaws and added sound/noises by scanning all parameters of sound at all levels of perspective.

Each device in the signal chain has potential to add noise to the signal, and to alter aspects of the signal in undesirable ways. A myriad of noises (such as clicks, hiss, hum, digital artifacts, quantization noise, and countless others) can be added to the signal by any device. They might also degrade the quality of the signal by introducing clipping, total harmonic distortion, phase shifting, drop out, intermodulation distortion, latency, and more. These can all be caused by misuse or malfunctions of the devices (i.e., a microphone overdriving a microphone preamp or a dirty potentiometer on a mixing console).

Interconnections of devices must also be correct to preserve the quality of signals throughout the signal chain. Gain staging, connection points, analog-to-digital and digital-to-analog conversions, conversions of digital formats, and many other issues also have the potential to diminish the integrity at any point throughout the signal chain. It is the engineer's responsibility to recognize all these events, their sources, and more to ensure a high-quality recording.

Other critical listening applications during recording and tracking sessions should be considered:

- Isolating sound sources during tracking, if separation is desired.
- Achieving desirable sound quality through matching an appropriate microphone with the sound source.

- Identifying an appropriate placement angle and distance for the microphone.
- Eliminating unwanted sounds created by performers or instruments.

Sounds will have a certain degree of isolation from one another. If the sounds are to be completely altered individually in the mixing stage, they must be isolated. When a group of sounds function as a single unit, it may not be appropriate to isolate the sounds from one another—they are often blended by the performance space. Problems arise when unwanted sounds leak onto tracks that contain sound sources that were intended to be isolated from all other sound sources. This leakage can be the cause of many problems later on in mixing directly to two-track or surround or in the multitrack mixdown process.

Sound quality should be carefully evaluated as it is captured by the microphone. The amount of timbral detail, blend, etc., of the sound sources will be determined now. These are major decisions that will have a decisive impact on the sound of the final recording and are largely determined by microphone selection and placement. Evaluations of the sound out of the context of the music is helpful in discovering subtle qualities that are often missed when one is listening analytically.

Some engineers rely on equalization and other processing to obtain a suitable sound quality during this initial recording. This use of processing alters the natural sound quality of the sound source and does not alter the timbre equally well throughout an instrument's range; for example, equalization will add a new formant region, resulting in an enhanced harmonic and altering the sound, of the track/sound source. It can be used effectively when the processed sound is desired over the source's natural sound or when practical considerations limit microphone selection and placement options. Often it can compromise the recording and should be approached carefully, in fact many engineers frown upon this activity stating that the best possible acoustic sound should be sought before engaging processing. As such, processing—especially equalization—can be used at this stage to compensate for poor microphone selection or placement.

Related to performance technique, the ways performers produce sound on their instrument, or the instruments themselves, can create unwanted sound qualities that must be negated during the tracking process. Often these aspects of sound quality are very subtle and go unnoticed until the mixing stage (when it is too late to correct them).

Instruments are capable of making unwanted sounds, as well as musical ones. The sound of a guitarist's left hand moving on the fingerboard, the breath sounds of a vocalist or wind player, and the "thud" from the release of a piano sustain pedal are but a few of the possible non-musical and (normally) unwanted noises that may be produced by instruments during the initial recording and tracking process.

These sounds are easily eliminated during the initial recording and tracking process through altering microphone placement, through slight modifications in performance technique, or through minor repairs to the instrument. The multitrack tape should be as free of all unwanted live performer sounds and sound alterations as possible. These sounds will be much more difficult to remove later in the recording process. They may be comprised of certain performance peculiarities that (depending on the situation) can only be alleviated by signal processing (such as the use of a de-esser on a vocalist).

Anticipating Mixdown During Tracking

The mixdown process must be anticipated while compiling basic tracks or indeed beforehand. It is necessary to determine how much control in combining sounds is needed or desired during the mixing process. Tracking, submixing, and isolation decisions can then be made accordingly.

Any mixing of microphones during the tracking process will greatly diminish the amount of independent control the engineer will have over the individual sound sources during the final mixdown process. Some mixing will often occur during the tracking stage, as submixes, to consolidate instruments and open tracks or to blend performers that must interact with one another for musical reasons.

Submixes will be carefully planned at the beginning of a session and must be executed with a clear idea of how the sounds will be present in the final mix. Drums are often condensed into submixes (either mixed live or through overdubbing and bouncing). Other mixes that will occur during the tracking process include the combining of several microphones (and/or a direct box) on the same instrument(s). The engineer and producer must be planning ahead to how these sounds will appear in the anticipated final mix—especially for musical balance, sound qualities, distance cues (definition of timbral detail), and pitch density.

Pre-processing (such as adding compression while recording) alters the timbre of the sound source before it reaches the mixing stages of the recording chain. Pre-processing also diminishes the amount of control the engineer will have over the sound during the mixing process. At times, it is desirable to pre-process signals; often it is not.

Desirable pre-processing might include stereo microphone techniques used on sound sources, effects that are integral parts of the sound quality of an instrument (distorted guitar), or processors that are used to provide a specific sound quality (compressed bass). At times pre-processing is used to eliminate unwanted sounds during tracking (such as noise gated drums). Once a source has been pre-processed, the alterations to the sound source cannot be undone. The engineer should be confident that they want the processed sound before recording it.

Some initial planning of the mixdown sessions will begin during the tracking process. Certain events that will need to take place during the mixdown session will become apparent as the tracking process unfolds. Keeping a record of these observations will save considerable time later on and may help other tracking decisions.

Examples of items that should be noted for the mixing process are:

- Sudden changes in the mix that may be required because of the content of the tracks.
- Certain processing techniques that are planned.

- Any spatial relationships or environmental characteristics that may be desired for certain tracks.
- Track noises or poor performances of certain sections that will need to be eliminated (muted) during the mix.

These are just a few examples of the many factors that may become apparent during the tracking process.

SIGNAL PROCESSING: SHIFTING OF PERSPECTIVE TO RESHAPE SOUNDS AND MUSIC

Signal processors are important tools (instruments) for the engineer. These devices may play a large role in shaping the individual project. Specific devices are chosen because their individual inherent sound qualities lend themselves to the particular project. They each control one of the three basic properties of the waveform—frequency, amplitude, or time, and thus the timbre overall.

Three types of signal processors exist, each functioning on a particular dimension of the waveform. An alteration in one of the physical dimensions of sound will cause a change in the other dimensions. Furthermore, alterations of the physical dimensions will cause changes in timbre (sound quality). The three types of processors do not only cause audible changes in the three characteristics of the waveform, they may also alter the timbre of the sound source. If considered according to how processors alter sound, signal processing can be simplified and approached with clarity.

- Frequency processors
- Dynamic processors
- Time processors

Frequency processors include equalizers and filters. Compressors, limiters, expanders, noise gates, and de-essers are essentially dynamic processors. Time processors are primarily delay and reverberation units. Effects devices are hybrids of one of these three primary categories. Some examples of these specialized signal processors include flangers, chorusing devices, distortion, fuzz, pitch shifters, and many more.

Uses of Processing

Signal processing can be used to shape sound qualities. It is applied to the sound source to complete the process of carefully crafting sounds. This is done for the character of the source's sound quality and to shape sounds to complement the functions and meanings of the musical materials and creative ideas.

In the recording chain, signal processing can occur at a number of times. It may be incorporated in the tracking as pre-processing and is most often used to bridge the tracking and mixdown processes. It can also be added subtly in mastering. Individual instruments or voices can be directed through any number of signal processors. Similarly, groups of instruments can receive the same processing, either in the same way or in differing amounts (such as any number of instruments each sending a different amount of signal to a buss feeding a reverb unit). The entire recording might also be processed, as is common in the mastering process.

Listening and Processing

During the production process, critical-listening decisions will be made to establish qualities of any sound, for its own sake. Decisions will also be made concerning how the sound relates to other sounds and to the entire program (analytical listening). The engineer will focus on the smallest changes in the source's timbre and on any number of higher levels of perspective. Careful attention to detail is needed, as well as attention to how small changes in one element or level of perspective impact the sound in other ways.

The engineer must use listening skills to focus on the component parts of the sound qualities of the sound sources being processed. Small, precise changes in sound quality are possible with signal processing. This requires the engineer to listen at the lowest levels of perspective and to continually shift focus between the various artistic elements (or perceived parameters) being altered. These changes are often subtle and can be unnoticeable to untrained listeners. Often beginning engineers are not able to detect low levels of processing. This is a skill that must be developed as introduced in Chapter 2A.

Most signal processing involves critical listening. The sound source is considered for its timbral qualities out of context and as a separate entity. In this way, the sound can be shaped to the precise sound qualities desired by the engineer, without the distractions of context.

Knowledge of the physical dimensions of the sound and perception is a great aid for successful signal processing.

Signal processing alters the electronic (analog or digital) representation of the sound source. In this state, the sound source exists in its physical dimensions of timbre (frequency, amplitude, and time). As such, the various signal processors are designed to perform specific alterations to the physical waveform, which will cause changes in the perceived timbre of the sound source. Signal processors are only useful as creative tools if the engineer is in control of these changes in the aforementioned physical dimensions. Immediately after altering the sound source's physical dimensions, the engineer can shift focus to place the sound in the context of the music, as an artistic element. The engineer will ask "Are these changes appropriate for the music, or achieve the desired sound for the musical instrument?" The sound quality shifts of processing will be evaluated according to their appropriateness to the musical idea.

THE MIX: COMPOSING AND PERFORMING THE RECORDING

Mixing is where the piece of music begins to emerge and ultimately comes together as a whole. The mix creates the piece of music, almost in its final form. Here the individual sound sources that were recorded or synthesized are combined into a two-channel or a surround sound recording that will become the final version of the piece after the mastering process.

It is often helpful to consider the mix process in two stages: one artistic, composing (crafting) the mix and one technical, performing (executing) the mix. While the two can happen almost simultaneously, they require different skills and thought processes.

The process of planning and shaping the mix is very similar to composing. Sounds are put together in particular ways to best suit the music. The mix is crafted through shaping the sound stage, through combining sound sources at certain dynamic levels, through structuring pitch density, and much more. How these many aspects come together provide the overall characteristics of the recording, as well as all its sonic details. Consistently shaping the "sound" of recordings in certain ways leads some engineers to develop their own personal, audible styles over time.

The actual process of executing the mix, as discussed in Chapter 5, is very similar to performing. Mixing often involves controlling sound in real time. Controlling the loudness levels of tracks, changing routing, muting tracks, altering processing, and more are routinely performed during mixdown—especially in systems that do not have automation. Many technical decisions also occur during mixing, many are out of real time. The engineer will develop their own approach to sequencing those activities and decisions, and ultimately create their own working methods.

Often, composing the mix and performing the mix happen simultaneously. The piece is shaped slowly as parts come together, and creative ideas are refined as new ideas emerge from hearing portions of the work as they are being completed. Sometimes new ideas take the project in new directions entirely, sometimes sounds or relationships of parts are "discovered" during the mixing process. Creative ideas get refined as a project unfolds. Keeping the technical process and all of the devices and controls from absorbing the creative energy and disrupting the flow of the project is necessary and is not an easy thing to accomplish. It is necessary for the engineer to learn to fluently operate all devices and software of the chain so they are in complete control of the production process. Only with the technical processes under control and in the background of the engineer's mind can attention truly be directed to crafting the artistic dimensions of the recording.

During the mixdown sessions the separate takes that were recorded during tracking and synthesis will be combined. Many of the recording's sound relationships are crafted in the process. To compose the mix, the engineer shapes the artistic elements. It can be helpful to group the elements into three groups or broad areas of focus. Approaching each separately in planning and executing the mix can help the engineer clarify their ideas and approaches to a mix. Each of these significantly shapes the recording, and each will also impact the other two. One must remember to listen to a mix from many perspectives and attention to all artistic elements, to be certain a desirable sound quality exists in all aspects of the recording.

These three groups are listed below, with the elements and some other concerns they contain:

1. Musical balance
   —Loudness
   —Prominence versus loudness
   —Attention, meaning, surprise

2. Pitch and sound quality
   —Pitch density and timbral balance
   —Performance intensity and sound quality
   —Environments of sources and sound quality
   —Dynamic contour by density and/or register versus loudness
3. Spatial qualities and relationships
   —Dimensions of the sound stage
   —Placement and size of sources on the sound stage
   —Listener to sound stage distance
   —Environments of sources and depth of sound stage.

It is important to remember the engineer must have or be working deliberately to establish a clear idea of the final sound qualities that are desired. With clear objectives, the engineer can work toward meeting those goals.

Successful mixes balance these elements in all of the sound sources with the message and musicality of the song. These elements are used to craft the overall characteristics of the song: form, structure, reference dynamic level, sound stage dimensions, perceived performance environment, timbral balance, and program dynamic contour.

Before we explore these three groups of elements individually, we will examine how the mix interacts with the music message and its musical materials.

Composing the Mix: Presenting, Shaping, and Enhancing Musical Materials

The song is made in the mix and the mix engineer is participating. The mix makes musical ideas come together. The song is built during the mixing process by combining the musical ideas, focusing on shaping the three groups of elements that will be discussed later.

A successful mix will be constructed with a returning focus on the materials of the song and the message of the music. The musical ideas that were captured in tracking are now presented in the mix in ways that best deliver the story of the text and the character of the music.

How the musical ideas are crafted and combined is in the area of music theory, and while not covered here this information is of great value to the engineer. The ways the recording process will combine the sound sources' materials will present their musical ideas in new relationships, and can profoundly shape and enhance them. The mix will greatly influence the musical style of the piece of music.

It is important to establish a clear idea of the final sound qualities that are desired for the song/recording. Just where to start will vary between individuals. Some people work from small ideas adding and building to create large sections. Some people need to have an idea of where they want to arrive, before they begin crafting the smaller details. Often individuals will use different approaches for different songs. How one arrives at a vision of what the song needs to sound like is not important. What is important is having a strong sense of the desired overall sound qualities of song, and a strong sense of how some (or most, or all) of the details of the music will bring this to reality.

The overall sound qualities, as explored previously, are:

- The song's sense of intimacy with the listener, created by the perceived listener to the nearest sound source distance.
- The energy level, intensity, and expressive qualities of the song that create a reference dynamic level.
- The overall dynamic shape of the song (program dynamic contour).
- How the song uses the pitch registers to create a timbral balance or spectrum of the overall recording.
- The width and depth of the sound stage (stereo or surround).
- The impression of the song's performance space or perceived performance environment.
- All of these and the musical structure coming together into a global shape and concept of the song, its form.

Some simple and direct questions can sometimes aid in determining or clarifying these areas:

- How does the story of the text unfold?
- How does the music support this?
- How can the mix support this?
- What relationship do we want the listener to have with the music (Observing from afar or intimately close? Maintaining a comfortable distance? In a large performance space or small? Focused on the text or feeling the beat? And many more.)
- What special qualities are needed to most effectively present what the song is trying to portray?
- How should each of the overall qualities contribute to communicating the music?
- Certainly many other questions are equally valid and potentially important.

The clearer the vision of what is sought in the areas, the smoother and more effectively the mix will progress.

Mixes create relationships of individual sources or small groups of sources when they are assembled. These sources (and the musical materials they are playing) are all given their own, or a shared "place" in the mix. This "place" might be a "space" or "location," a "level" or "area," a "character" or a "set of characteristics" in each artistic element—such as lateral location, register, dynamic level, etc. In this way, mixing is the act of putting everything in its place, while giving any final shapes to the sounds.

Just where to place a sound source is a matter of what will best suit the song and what will deliver the desired overall sound qualities. Among important questions to ask are:

- How can this instrument/voice, presenting this musical idea be placed in the musical balance to contribute most effectively?

- What spatial qualities and relationships will most effectively present this instrument/voice and its musical material?
- What sound qualities are best suited to this instrument/voice presenting this musical idea?
- How can I best present or enhance this sound source and musical idea?
- Should this idea and instrument (bass, for instance) be emphasized, or should it be blended with others (bass, keyboard, and bass drum)?
- What special qualities do we want to bring to this instrument to enhance the song?
- How can these instruments be combined to provide the sound qualities and sound stage that is desired?
- What musical materials/instruments contribute most to defining the energy or the character or the message or the sound quality or the expression of the piece—and how should they be treated?
- What musical materials are supporting the primary ideas, and how can they be made to do this most effectively?

The answers to these questions (and others that are similar and others that are more detailed later in Chapter 5) will lead the engineer with direction and purpose to crafting a mix that supports the music and presents it in the most appropriate way.

The engineer will decide when to blend sources together and when to allow a source to be prominent. Just as importantly they will decide on how the characteristics of sound will allow these to happen. We remember that sounds may have any level of importance to the mix and have many dimensions. Sounds can be blended with others into groups or be clearly delineated from others by unique qualities. All of the elements can give coherence to the mix by providing groups of sources with similar qualities and can provide variety to the mix by providing sources with unique qualities. Different elements will function differently on the sound sources, such as some instruments might be grouped by pitch density, as they perform in a similar pitch range, but be delineated by a different stereo location. The combinations and subtle variations are about limitless and provide the basis for shaping the artistic qualities of the music. Without a clear idea of how the finished mix is intended to sound, the engineer will have great difficulty in making quality decisions.

Stating the seemingly obvious, songs have sections with different musical materials, and often the sections contain different instrumentation; songs change from beginning to end in many possible ways, sometimes markedly and sometimes subtly. Thinking in such simple terms can be of great assistance in compiling mixes, especially in the beginning. Mixes have the potential to change throughout a song as well and this is not unusual. Several distinct mixes might be used in a song, changing with the song's various sections (such as verse or chorus) or within sections. Mixes can change in an unlimited number of ways, just like the musical materials of songs. Some of these changes might be very subtle and some might be striking. A few potential large-scale changes might be:

- marked changes between sections (such as verse and chorus);
- subtle changes over the course of the song;

- marked changes of only a few sound sources (such as lead vocal and drum set) between sections;
- sudden entrances of large groups of instruments with corresponding changes in the mix;
- sudden exits of nearly all instruments to reveal only a few sources with corresponding changes in the mix for the remaining sources.

Referring to the three concepts to be explored below, mixes might be changed in all or one of the following groups of elements:

- changing relationships of musical balance of some or all sound sources;
- changing spatial qualities and relationships of some or all sound sources;
- changing pitch and sound qualities of some or all sound sources.

The mixing process taxes listening skills. Acute attention to all elements of analytical listening and critical listening are vital. While the perspective is usually at the individual sound source and up to one level to how the sources sound against one another, strong awareness to the overall texture is required to shape those characteristics discussed above. Close attention to the detailed characteristics of sources, to the integrity of the signal, and the technical qualities of the recording are also required. The engineer will be continually engaged in shifting focus from one element to another and from one level of perspective to another. The engineer must continue to seek greater detail and an awareness of subtle qualities and relationships of sound sources and materials: the subtle characteristics of a recording often separate the remarkable mix that presents the music in an engaging way and from one that is acceptable but uninspiring, or worse from one that is poor or ineffective.

Crafting Musical Balance

The entire mixdown process is often envisioned as the process of determining the dynamic level relationships of the sound sources. As we have noted, the mixdown process is actually much more. It combines many complex sound relationships, of which dynamics is only one—but an important one.

Sound sources in the mix are related to one another by dynamic level, just as in live performance. The difference is that the levels can be carefully calculated and changed throughout the course of the mixdown process. The mixing console or mix function of a DAW allows the engineer to perform the individual dynamic levels of the sound sources and to make any changes in level in real time. This significantly shapes the mix and the music.

The engineer devises a musical balance of the individual tracks and performers. This is the relationship of the dynamic levels of each instrument to one another and to the overall musical texture. The individual sound sources are combined into a single musical texture, each source at its own loudness level. Small changes in level of a single sound source can be difficult to detect, especially in the beginning; this is changing the dynamic contour of a line, and hearing these changes is an important skill to be developed. Dynamic relationships are more readily perceived at a higher level of perspective—one that compares sources to one another in the overall texture; beginning engineers can more readily develop the skill of hearing these relationships earlier in their work. Obtaining the ability to recognize and control loudness changes and relationships is necessary. A slight alteration of musical balance and/or the dynamic shape of a musical idea can have a profound impact on the music.

Changing loudness levels in a mix readily creates a difference between the actual loudness of the sound source and the performance intensity at which it was performed during tracking. Sound quality and loudness are then separated. This difference between musical balance and performance intensity skews the impression of a live performance and interaction of the musicians. Potential for dramatic and creative dimensions in altering the realities of sound quality and dynamic level relationships exists, or this can cause a desired live-sounding recording to be distorted.

With sound source loudness levels carrying characteristic timbres, a psychological effect exists whereby sounds might be imagined to be at the loudness level of their performance intensity, when actual loudness is very different. A much greater imagined loudness can occur when a sound is recorded moving from very soft to very loud, while a compressor holds loudness almost constant at a low level. Here the timbre of the sound source, and our knowledge of the amount of energy and expression placed into the performance, brings an impression of a higher loudness level, although the actual loudness (amplitude) is markedly different. This effect can be used to creative advantage to give prominence to a sound without increasing its loudness.

Perception of actual loudness level is often mistaken with other things. It is often easy to confuse the prominence of a musical idea or instrument with loudness. A sound may be prominent because of some special quality (register placement, for instance) while actually being at a lower loudness level than other sounds. A sound can be the most prominent in the listener's consciousness while being at a lower dynamic level than other sounds. The other artistic elements have equal potential to provide outstanding qualities to the sound and to cause the sound to stand out from the musical texture.

In similar ways, loudness is often confused with or distorted by the listener's attention to certain aspects of a song, by unexpected events in the music and by the meaning of a text or the music. The listener may be drawn to the text of a song, and the singer might be perceived as the loudest musical part; while this can be the case, it often is not. This understanding can sometimes allow the engineer to lower the loudness level of the vocal without moving the attention of the listener.

When something new or unexpected happens in a song the listener often shifts attention to it. This can cause the listener to perceive the event as louder than it actually is. For instance, one might incorrectly perceive the high hat sound in the second verse of "Let It Be" by the Beatles (Let It Be version) to be louder than the voice. In closer listening it is clearly softer. This misperception is caused simply because the high hat arrival is attention getting and a surprise since no percussion sounds have preceded it—and also because the high hat's environment pulls it across the sound stage and because it is in a very different pitch area than the lead vocal and piano.

Pitch and Sound Quality Concerns in Combining Sound Sources

The mix also combines the sound qualities of sound sources. When sounds are combined in the mixing process, the timbres of the instruments/voices are blended. This blending of sound source timbres can bring sounds to fuse together into a group, or if handled differently the sound sources can retain all or some of their unique characters even if they occupy very similar timbres pitch area. Carefully employing the other elements of sound (such as stereo location, loudness, etc.) can keep similar timbres from fusing in the mix.

The sound quality of the sound source plays a significant role in the successful presentation of the musical idea. The instrument or voice's timbre is shaped to most effectively present the musical material, and the listener will ultimately come to identify the musical idea by the instrument (or singer) that delivers the musical idea. This process of shaping the sound quality of the sources began in tracking, with instrument selection, microphone selection and placement, performance intensity and expressive qualities of the performance, and perhaps signal processing. Sound sources will now receive final shaping during mixdown by signal processing (whether hardware or plug-ins).

This final shaping can be used to enhance the character of the sound, or to help a sound combine more effectively in the mix. Time, spectrum, and amplitude processes can all be employed for these purposes.

The sound quality of sound sources also contains the dimension of environmental characteristics. We recall the sound source and its host environment fuse into a single impression. In this way, the sound qualities of the environment become a part of the sound qualities of the sound source. Shaping of environmental characteristics must, therefore, be viewed from the perspective of how they impact the sound quality of the source. This can have a significant impact on the source's pitch area.

A source's sound qualities may remain constant throughout the piece, or the qualities may make sudden changes or be gradually altered in real time during the mix. Many possibilities exist for shaping and controlling sound quality.

When instruments and voices get placed in the mix, they should be evaluated for their frequency content. This is comprised of the pitch(es) they perform plus the spectrum of their timbre (including environmental characteristics). In this way, all sounds occupy an "area" in the frequency range of our hearing. This is a bandwidth, of sorts, where the instrument's sound and musical material combine and occupy. This is the pitch density of the musical idea. The pitch densities of all of the sound sources combine to create the timbral balance or the "spectrum" of the overall texture.

It is common for some types of music to emphasize certain pitch registers over others. For example, the rhythm section of the typical rock band will have the bulk of its pitch-plus-timbre information (or pitch area) in the "Low," "Low-Mid," and "Mid" ranges. The timbral balance is weighted in these low frequency ranges, and instruments and voices performing above these registers occupy a very different pitch area and are easily perceived even when performing at lower dynamic levels or placed in the same locations on the sound stage. Thus, when sounds are combined, a source's pitch area can be exploited to bring the source to blend with others or to be more readily perceived.

As this song progresses, the rhythm section might thin out at times or perhaps stop. This causes a shift in the timbral balance. Such shifts are common in music. Shifts of timbral balance often happen with changes in the mix, at climactic points in the music, with the entry and exiting of instruments/voices, between a verse and chorus, and more. Shifts are common; some are very subtle and some are striking. This emphasis of some pitch registers over others provides an overall sound quality to the song that comprises its timbral balance.

Timbral balance can contribute to shaping the program dynamic contour of the work. The density of pitch areas and register placement can increase or decrease the overall loudness of the program. In this way, the dynamic level of the overall program is shaped by the number of sound sources present and the registers in which they sound as much as the loudness levels of the sources. This is an important consideration when adding instruments and voices to the mix, or pulling sounds out.

Crafting sound qualities and creating pitch densities and a timbral balance are new forms of arranging and orchestration. The engineer will be required to listening in many different ways to make these important decisions. They will at times focus on the dimensions of the individual sound qualities. When the sound qualities of the sources are combined to create the sound qualities of groups of instruments and the overall sound quality of the music (timbral balance), the focus of the engineer will shift perspective between these various levels while continuing to scan between the components of each source's timbre.

Throughout the process of compiling (composing) the mix, the engineer's attention will return to timbre and sound quality. This is accomplished using both analytical and critical listening skills. Timbres will be considered as separate entities (out of time) and as sound quality in the musical contexts of all hierarchical levels.

Creating a Performance Space for the Music and the Recording

Sound sources are given spatial qualities during mixing. These qualities provide an illusion of a performance space for the recording—An imaginary place where the music was performed, with dimensions and sound qualities. The mix creates these dimensions and qualities.

Individual sources are shaped in terms of placement on the sound stage and environmental characteristics. Sounds are placed on the sound stage at specific locations and are placed in environments.

Spatial properties for the overall program are also crafted in the mix. The spatial properties created in the mix provide an illusion of a space within which the performance takes place. As sounds are placed at locations within this perceived performance environment, the dimensions of the sound stage are defined.

The spatial qualities that are crafted during mixing are:

- Dimensions of the sound stage
- Placement and size of sources on the sound stage (lateral and distance)

- Listener to sound stage distance
- Environments of sources and depth of sound stage
- Perceived performance environment.

During the mix, sounds will be placed at specific lateral locations, as phantom images or at speaker locations. These images will be given a width anywhere from the breadth of the entire available sound stage to a narrow point in space. This lateral location can have great impact on separating sounds in the mix or blending them. The size and location of sounds can provide prominence or importance for a sound that would otherwise be less noticeable.

It is helpful to remember humans do not localize sounds equally well at all frequencies. Further, we primarily use interaural time information to localize sounds below 2 kHz and interaural amplitude cues above 4 kHz to localize sounds. This requires the engineer to consider the pitch area of the sound source, and to work with amplitude, time/phase, and spectrum appropriately to accurately create stable images.

Taken as a sum, the lateral placements of all of the sound sources provide the listener with a sense of width of the sound stage. The listener develops a sense of "where" the sounds are and the size and location of the stage where the song is being performed. This shapes an overall quality of the recording that is very important.

The location and size of the sound stage has become increasingly significant, as the industry has engaged surround sound production practice. This is a major factor that separates different approaches to surround sound production and revolves around how the rear channels are used for different types of program materials (especially musical materials and reverberant sound). When sound sources are placed behind the listener, for instance, they might perceive themselves as sitting within the sound stage. The rear channels might also be used to pull the sides of the sound stage wider than is possible with two-channel stereo playback. Of course, it is also possible (and not uncommon) for the sound stage to remain in front of the listener, and the lateral imaging be very similar to a traditional stereo recording, with the addition of manufactured ambient sound appearing behind the listener.

The other dimension of the sound stage is distance. Sounds are placed at a distance from the listener and provide the illusion of depth to the recording. As a group, the distances of all of the sound sources provide the front-to-back dimension of the sound stage.

Distance placement can provide a special quality to a sound source. It can allow a sound to be clearly apparent in a musical texture or to blend more with other sounds. The importance of distance cues is often underestimated. These cues bring musical materials and instruments/voices into a physical relationship to the listener that can be profoundly effective in helping the musical message or expressive nature of a line. Distance can bring a musical idea to an immediacy to the listener or can provide a sense of being removed from the source and musical idea; this has the potential to greatly shape the listener's sense of the musical idea. A very different sound stage exists when all sources are at approximately the same distance, as they appear to be performing in a similar area and have a sense of connection in space, than when all sound sources are at even slightly different distance relationships, as they appear to extend the sound stage and bring sources to have differing relationships to the listener. In recordings where sources are extended from close proximity into the farther areas, with many sounds in between, the sound stage can achieve vast proportions and bring a substantial new dimension to a mix.

The stage to listener distance is critically important to establishing the level of intimacy of the recording. The song can speak intimately to an individual listener when the lead vocal is very close and clearly within their sense of proximity. The song can have a different character with a moderate distance, which places the listener in the position of observing what is being said by the song instead of being personally engaged in it by a close distance relationship.

It is important to remember that distance cues are primarily the result of timbral detail. A high degree of low-level detail must be present for a sound to appear very close. Sounds become more distant as this detail is removed. Placing a sound in the mix can alter its timbral detail, as it blends with or is masked by other sounds. Being sensitive to pitch density and lateral location are especially helpful in preserving a sound's timbral detail. Reverberation can alter distance. This can be because of the ratio of direct to reverberant sound but is often because the reverberant sound masks the timbral detail of the direct sound. The reverberation's arrival time gap and the reflections of the early sound field will also provide subtle cues that can shape the imagined distance of the sound source. When adding reverb to a signal it is important to bring attention to how it is impacting distance.

The listening perspective of the engineer will alternate between locating individual sound sources on the sound stage, comparing locations of sources to one another, observing how all sources create an overall sound stage (and observing how balanced the stage might be), and recognizing how the sound qualities of sources might be transformed by placing the sounds on the sound stage. These observations will be made for both distance and lateral location cues.

Sound stages can be planned and evaluated using Figures 3.1 and 3.2. These will prove helpful in balancing the sound stage and in creating variety and interest—as desired—of image locations and distances. These figures are snapshots of time and may represent any time unit from a moment to a complete song. They allow imaging to be recognized and understood. The dimensions of the sound stage can be drawn around the loudspeakers in each figure, and the listener location will be determined in two-channel recordings. The figures should show the front edge of the sound stage, giving the engineer a reminder of the sense of intimacy of the recording, and the depth of the sound stage providing important environment size and distance information. A significant set of sound relationships can be planned, crafted, and evaluated with these diagrams.

Each individual sound source is placed in an environment during the process of making the recording. Qualities of the recording environment may have been captured during tracking, and these may provide all of the environmental characteristics that are desired. Very often environmental characteristics are added to a sound source or a group of sound sources during the mix. If a sound is added into the mix

FIGURE 3.1
Sound stage diagram
for two-channel
recordings.

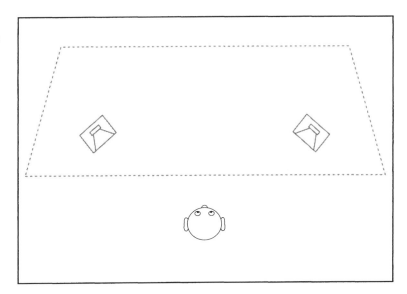

FIGURE 3.2
Sound stage diagram
for surround sound
recordings.

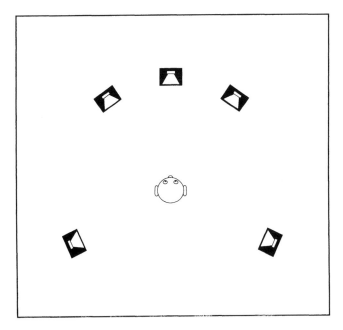

without environmental characteristics (such as a direct-in electric bass) the listener will imagine an environment (often a very small one).

Environments have sound qualities. When a source is placed in an environment, it acquires the spectrum and time/reflection components of the environment. The environment and sound source timbres are fused into a single and new sound quality.

Environmental characteristics can add important dimensions to a sound source and shape their sound quality in significant ways. They also can add to the dimensions of the sound stage. The environment of a sound source can extend the rear of the sound stage, especially with long, high-level decays in the reverberation. The perceived size of the environment of a sound source can shape its musical material and also shape the sound stage itself.

In surround sound production, it sometimes happens that the environmental characteristics of a sound source are separated from sources themselves. As such, the direct sound appears in one location and the reverberant information in another. Since the listener naturally fuses the two sound qualities, they are now clearly distinguishable because of different locations, a new relationship is achieved. This is different from separating the ambience of the entire mix with a sound stage located elsewhere. Careful crafting of the mix will be required to allow this new relationship to be acceptable to the listener.

Finally, the engineer must turn attention to shaping the environmental characteristics of the overall program. The sound stage will be heard as existing within a single performance space, the perceived performance environment. The dimensions of the perceived performance environment may be applied during the mastering process. A subtle reverberation program can be applied to the final mix to provide an environment for the recording. In this case, the engineer should plan for this change in sound quality while compiling the mix, paying special attention to timbral balance and distance cues.

Often the environments of several or all individual sound sources will create the perceived performance environment. In this case, elements of the environmental characteristics of some or all sound sources are heard as important to the character of the work and become perceived as dimensions of the perceived performance environment. No environment is added to the final mix, but the listener imagines one. Usually, the most important instruments and voices shape these elements but unusual environments of lesser important sources can have a strong impact. In this way, the shaping of the environmental characteristics of the sound sources that present the most important musical materials will have a direct and marked impact on the listener's impression of the performance environment within which the recording itself appears to take place. The engineer must be aware that how certain individual environments are crafted could impact this overall characteristic of the recording.

Shaping environmental characteristics requires focused attention at the lowest levels of perspective for spectrum, spectral envelope, amplitude, and time information. Focus will shift to a higher level to evaluate how the characteristics of the environment alter the source's sound quality and will ultimately move to the highest level to observe the environment of the overall program.

Crafting spatial properties is as important to the mix as any other element. They may be used to delineate the sound sources into having their own unique characteristics, or they may be used to cause a group of sound sources to blend into a sense of an ensemble. It is possible for a group of instruments to be grouped in several spatial dimensions (such as an environment) but to have very different characteristics in another (such as distance). Common characteristics provide a connection between sources and unifying them in some way. Differences distinguish sound sources and can add prominence to their presence in the mix and perhaps call more attention to their musical material.

Performing the Mix

In performing the mix, the sound qualities and relationships of the recording are realized. How everything comes together depends on technologies and working methods.

The process of executing the mix is similar to performing, in that sound is being controlled, often in real time. Many technical decisions also occur during mixing, many of which are out of real time. These activities are planned, just as sound qualities and relationships were planned while "composing" the mix. Details pertaining to specific recording equipment, technologies, and techniques will directly impact how the mix is planned.

Work methods will vary by individuals, but they will all center on establishing a logical, efficient, and effective sequence of activities. The goal is to best use the selected equipment and technologies to craft the mix as planned and to establish and maintain a high-quality signal. Equipment and technologies vary greatly, but the process remains the same; the basic signal chain and the events of the production sequence remain constant and should remain clearly in mind. Once the engineer has obtained knowledge of technologies and mastery of software and equipment usage they can develop their own approach to sequencing production activities and decisions, and might ultimately create their own working methods. As these methods are technology and equipment dependent, they are outside the scope of this writing, but further reading lists can be found throughout this book.

Recording aesthetics and the planned qualities of the recording will shape the production process. Many possibilities for approaching the mix exist and the correct approach for one project is not necessarily correct for another. For example, multitrack mixes are often "performed" a track or two at a time but sometimes any number of tracks are crafted simultaneously; in direct-to-master recordings all instruments are performed simultaneously. In another example, sounds might be placed in the mix before sound qualities receive their final shaping (with final EQ or reverberation added later) in one project, but frequently sound qualities are finalized well before sources are placed in the mix and combined with other sounds.

Skill in performing the mix relies on knowledge of the technologies being used in the signal chain, on mastery of equipment and software usage and the interface points of the signal chain, and on dexterity with the techniques of creatively using these devices and technologies to craft the mix.

SUMMARY

The mix is the recording in nearly its final form and is where the piece of music comes together. A vision of the sound qualities and relationships of sounds in the recording must be present at the start of the recording process. Preparation for the mix begins as far back as selecting sound sources suitable for the musical materials and through to capturing or creating the performances of the sound sources during tracking or recording sessions.

The mix is composed through planning the qualities and relationships of sounds and the overall characteristics of the recording. Details about recording equipment, technologies, and recording techniques will achieve these qualities and relationships can also be planned. During the mix the sound qualities, dynamics, and spatial qualities of the sound sources and the recording are then carefully shaped.

The sound sources may be recorded in a multitrack format (in a DAW or to analog multitrack tape, for example), for mixing at a later time, or mixed directly to the recording's final format (typically two-channel stereo or 5.1 surround).

The mix results in the final version of the piece of music. This recording will often be transformed a final time in making the master of the recording.

FURTHER READING

This is an edited chapter from Moylan, W. (2007). *Understanding and Crafting the Mix: The Art of Recording*. Focal Press, Oxford.

Hatschek, K. (2005). *The Golden Moment: Recording Secrets from the Pros*. Backbeat Books.

Moulton, D. (2000). *Total Recording: The Complete Guide to Audio Production and Engineering*. KIQ Production, Inc., Sherman Oaks, CA.

SECTION B
The Sessions

Chapter 4
The Recording Session

David Miles Huber

In This Chapter

INTRODUCTION

One of the most important concepts to be gained from books such as this is the fact that there are no rules for the process of recording music. This rule holds true insofar as inventiveness and freshness tend to play a major role in keeping the creative process of making music (and music productions) alive and exciting. There are, however, guidelines and procedures that can help you have a smoother flowing, professional sounding recording session or, at the very least, help you solve potential problems when used in conjunction with three of the best tools for guiding you toward a successful project:

1. Preparation
2. Creative insight
3. Common sense

PREPARATION

As we discussed in Chapter 2, probably the most important step that one can make to help ensure that a recording project will be successful and that it has a chance at being marketable to its intended audience is careful *preparation* and *planning*. By far the biggest mistake that a musician or group can make is to go into the studio, spend a lot of money and time, press a few thousand CDs, make a template website and then sit back and expect an adoring audience to spring out of thin air! It ain't gonna happen! Beyond a good dose of business reality and added experiences, the artist(s) will have the dubious distinction of joining the throngs that have thousands of CDs sitting in their closet or basement.

In order to help avert many of the more common mistakes, the first half of this chapter is devoted to laying out many of the tools that can be gathered to help ensure that a project will go smoothly, be on budget, sound good, and, of course, sell. The latter half will be devoted to the actual process of recording in a professional or project studio. To those that choose to dive into the deep end of music and sound production (on either side of the glass)—all the best!

What's a Producer and When Do You Need One?

One of the first steps that can help ensure the success of a project is to seek the advice and expertise of those who have experience in their chosen fields. This might include seeking legal council (for help and advice with legal matters, business contacts, or both). Another important "advisor" can come in the form of that all-important title, *producer*. Basically, the producer of a project can fill one of two roles:

- The first type can be likened to a film director, in that his/her role is to be an artistic, psychological, and technical guide that will help the band or artist reach their intended goals of obtaining the best possible song, album, remix, film score, etc. It is the producer's job to stand back and objectively look at the big picture, and to offer up suggestions as to how to shape and guide the performance and to direct the artist or group in directions that will result in the best possible final product.
- The second type also encompasses the directorial role, but also has the added responsibilities of being an executive producer. He or she will also be charged with many of the business responsibilities of overall session budgeting, choosing and arranging for all studio and session activities, contracting (should outside musicians and arrangers be needed on the project), etc. This type of producer may even be charged with initiating and/or smoothing contact relations with potential record companies or distributors.

As you can see, this role can be either limited or broad in scope and should be carefully discussed and agreed upon long before any record button is pressed. The importance of finding a producer that can work best with your particular personalities, musical style, and business/marketing needs can't be stressed enough. Finding the right producer for you can be a time-consuming and rewarding experience. Here are a few tips to prepare you for the hunt:

- Check out the liner notes of groups or musicians that you love and admire. You never know, their producer just might be interested in taking you on!
- Find a local up-and-coming producer that might be right for your music. This could help fast track your reputation.
- Talk with other groups or musicians. They might be able to recommend someone.

A few of the questions to ask when searching out a producer are:

- Does he/she openly discuss ideas and alternate paths that contribute to growth and better artistic expression?
- Is he/she a team player or are the rules laid out in a dictator-like fashion?
- Does the producer know the difference between a creative endeavor and one that wastes time in the studio?
- Does he/she say "Why not?" a lot more often than "Why?"

Although many engineers have spent most of their lives with their ears wide open and have gained a great deal of musical, production, and in-studio experience, it's generally not a good idea to assume that the engineer can fill the role of a producer. For starters, he/she will probably be unfamiliar with the group and its sound, or may not even like your sound! For these and other reasons, it's always best to seek out a producer that is familiar with you, your goals, and your style (or is contacted early enough the he/she has time to become familiar).

Long Before Going into the Studio

Most in the industry realize that music in the modern world is a business. Once you get to the phase of getting your band or your client's band out to the buying public, you'll quickly realize just how true this is. Building and maintaining an audience with an appetite for your product can easily be a full-time business; one where you'll encounter well-intentioned people, and also others who would think nothing of taking advantage of you or your client.

Whether you're selling your products on the street, at gigs, or over the Internet, or whether you're shopping for (or have) a record label, it's often a wise decision to retain the counsel of a trusted music lawyer. The music industry is fraught with its own special legal and financial language, and having someone on your side who has insight into the language, quirks, and inner workings of this unique business can be an extremely valuable asset.

By far, one of the most important steps to be taken when approaching a project that involves a number of creative and business stages, decisions, and risks is preparation. Without a doubt, the best way to avoid pitfalls and to help get you, your client, or your band's project off the ground is to discuss and outline the many factors and decisions that will affect the creation and outcome of that all-important "final product." Just for starters, a number of basic questions need to be asked long before anyone presses the "REC" button:

- How are we going to recoup the production costs?
- How is it to be distributed to the public? Self-distribution? Indie? Record company?
- Will other musicians be involved?
- Do we need a producer or will we self-produce?
- How much practice will we need? Where and when?
- Should we record it in the drummer's project studio or at a commercial studio?
- If we use the project studio and it works out, should we mix it at the commercial studio?
- Who's going to keep track of the time and budget? Is that the producer's job? Or will he/she be strictly in charge of creative and contact decisions?
- Are we going to need a music lawyer with contacts and contracts? Do we know someone who can handle the job?
- Carefully discuss the artist or group's artistic and financial goals and put them down on paper. Discuss budget requirements and possible rewards as early as possible in the game! This might be the time to discuss matters with the music lawyer we talked about.

These are but a few of the questions that should be asked before tackling a project. Of course, they'll change from project to project and will depend on the final project's scope and purpose. However, in the final analysis, asking the right questions (or finding someone who can help you ask the right questions) can help keep you from having to store 10,000 unsold CDs in your basement.

Now that you've answered the questions, here's a list of tasks that are often wise to tackle well before going into the studio:

- Practice, practice, and more practice. Need I say more!
- Create a "Mission Statement," perhaps using the V.I.S.I.O.N. method outlined in Chapter 2A as a basis, for you/your group and the project. This can help clue your audience into what you are trying to communicate through your art and music and can greatly benefit your marketing goals. For example, you might want to answer such questions as: Who are you? What are your musical goals? How should the finished project sound? What emotions should it evoke? What is the budget for this project? How will it be sold? What are the marketing strategies?
- Start working on the project's artwork, packaging, and website ASAP.
- Copyright your songs. In the United States, Form PA is used for the registration of music and/or lyrics (as well as other works of the performing arts), while Form SR is used to copyright the actual recorded performance. These forms can be found at www.copyright.gov/forms or by searching the Library of Congress at www.loc.gov.
- Take the time to check out several studios and available engineers. Which one best fits your style, budget, and level of professionalism?

Before Going into the Studio

In Chapter 2 we introduced the notion of preparing for the process as a whole. Here's a really useful list of things to consider discussing with the production team before walking into the studio.

Before beginning the recording session (possibly a week or more before), it's always a good idea for the engineer to mentally prepare for what lies ahead. It's always a good idea to pass on a basic checklist that can help answer what type of equipment will be needed, the number and type of musicians/instruments you'll be bringing, their preferred particular miking technique (if any), and where they like to be placed (next to the drummer for example). The best way to do this is for you, your group, and the producer (if there is one) to sit down with the engineer and discuss instrumentation, studio layout (Figure 4.1), musical styles, production techniques, and the vision for the project. This meeting lets everyone know what to expect during the session and lets everyone become familiar with the engineer, studio, and staff. This is time well spent, as it will invariably come in handy during the studio setup and will help get the session off to a good start. The following tips can also be immensely valuable:

- If there is no producer on the project, it's often wise to pick (or at least, consider picking) a spokesman for the group who has the best production

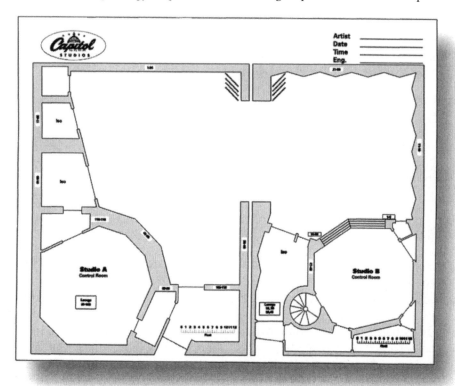

FIGURE 4.1
Floor plan layout for Capitol Studios A and B (which can be opened up to create a shared space). Courtesy of Capitol Studios, Hollywood, CA, www.capitolstudios.com.

"chops." He/she can then work closely with the engineer to create the best possible recordings.

- As discussed in Chapter 2, you should look to record your songs during rehearsals. It doesn't matter if you record them professionally, with a 4-track, or with a cassette recorder/boom box. Another great alternative is to record your live gigs as this will also show how the audience reacts to certain songs.
- Audition the session's song list before a live audience.
- Work out all of the musical and vocal parts before going into the studio. Unrehearsed music can leave the music standing on shaky ground; however, leave yourself open to exploring new avenues and surprises that can be the lifeblood of a magical session.
- Rehearse more songs than you plan to record. You never know which songs will end up sounding best or will have the strongest impact.
- Meet with the engineer beforehand. Take time for the producer and/or group to get to know him/her, so you'll both know what to expect on the day of the session. You might ask to hear examples of what the engineer has recorded.
- Prepare and edit any sequenced, sampled, or pre-recorded material beforehand. In short, be as prepared as possible.
- Try working to a metronome (click track) if timing is an issue.
- Take care of your body. Try to relax and get enough sleep before and during the session. Eat the foods that are best for you (you might have some health foods or fruits around to keep your energy up). Don't fatigue your ears before a session, keep them rested and clear.

In addition to the above, it's always a wise idea to discuss with the production team how they plan to arrange the tracking. If you're using tape-based equipment, decide on your tracking order and make sure that there are enough tracks for all of the instruments (with a few to spare). Confer to find out how many instruments are to be used on the song, including overdubs. This helps to determine how many tracks will need to be left open. The number of tracks often influences the number and way that mics are to be assigned to the available tracks. This holds especially true for drums.

If a large number of instruments are to be recorded and overdubbed to a recorder with a limited number of tracks, you might want to consider the following:

- Group several instruments together and record the composite mix to a limited number of tracks.
- Record the instruments onto separate tracks and then mix the composite mix (or mixes) onto a small group of tracks on a separate master tape leaving tracks open for additional overdubs. Keeping the original tracks intact on a separate tape will at least give you more options should the group mixes need to be changed at a later date (see the Bouncing section on page 89).
- Decide to use (or rent) a recorder that has more tracks or integrate one or two additional modular digital multitracks into the recording setup. In this day and age, there are often several options for adding on tracks.
- Use a hard-disk based digital audio workstation (DAW) that has fewer track limitations.

SETTING UP WITH THE PRODUCTION TEAM

Once all the musicians have shown up at the studio, it's extremely important that all of the technical, musical, and emotional preparation be put into practice in order to get the session off to a good start. Here are a few tips that can help:

- The engineer should show up at the studio on time or much earlier, as should the musicians. At some studios, the billing clock starts on time (whether anyone is there or not). Ask about their setup policy. (Is there another session before yours? Is there adequate setup time to get prepared? Are there any charges for setup? What is the studio's cancellation policy?)
- Use new strings, drumsticks, and heads, and bring spares. It is also a good idea to know the location and hours of a local music store, just in case.
- Tune up before the session starts, and tune up regularly thereafter.
- Don't use new or unfamiliar equipment. Taking the time to troubleshoot or become familiar with new equipment can cost you time and money. The frustration could even result in a lost vibe! If you must use a new toy or tool, it's best to learn it beforehand.
- Have the producer and/or group go over the song list, arrangements and any other planned information with the engineer. This might help smooth out setup problems and help the engineer to better understand the goals of the session.
- Take the time to make the studio a comfortable place in which to work. You might want to adjust the lighting to match your mood ring, lay down a favorite rug and/or bean bag, turn on your lava love light, or put your favorite stuffed toys on the furniture. Within reason, the place is yours to have fun with!

From a technical standpoint, the microphones for each instrument are selected either by experience or by experimentation and are then connected to the desired console inputs. The input used and mic type should be noted on a track sheet or piece of paper for easy input and track assignment in the studio and/or for choosing the same mic during a subsequent overdub.

Some engineers find it convenient to standardize on a system that uses the same mic input and tape track for an instrument type at every session. For example, an engineer might consistently plug the kick drum mic into input #1 and record it onto track #1, the snare mic onto #2, and so on. That way, the engineer instinctively knows which track belongs to a particular instrument without having to think much about it.

Once the instruments, baffles, and mics have been roughly placed and headphones that are equipped with enough extra cord to allow free movement have been distributed to each player, the engineer can now get down to business by using his/her setup sheet to label each input strip with the name of the corresponding instrument. Label strips (which are often provided just below each channel input fader) can be marked with an erasable felt marker, or china graph pencils, or the age-old tactic of rolling out and marking on a strip of paper masking tape could be used (ideally, a kind that doesn't leave a tacky residue on the console surface).

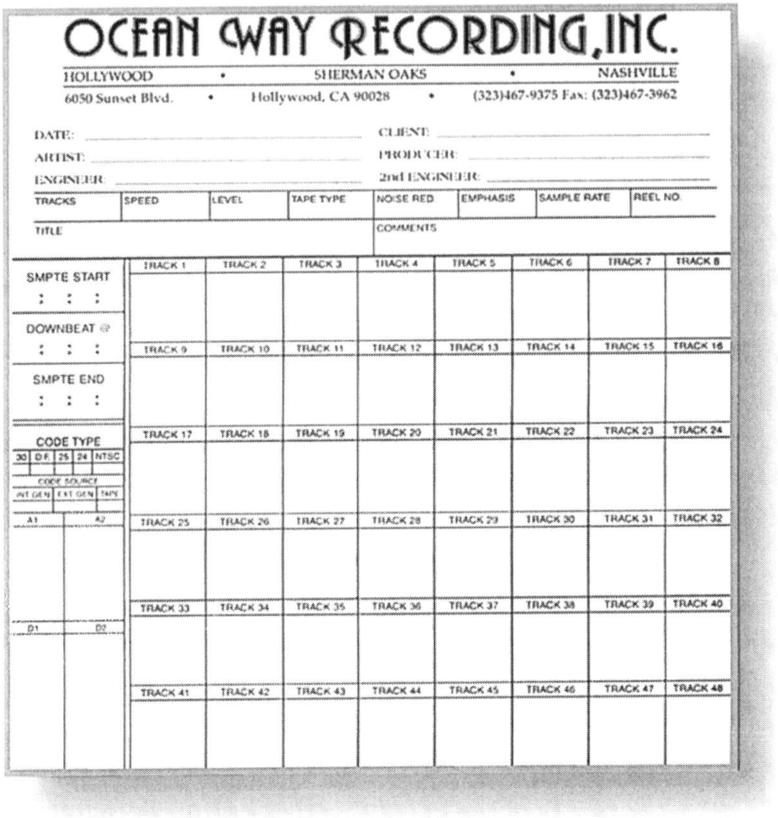
OCEAN WAY RECORDING, INC.
HOLLYWOOD • SHERMAN OAKS • NASHVILLE
6050 Sunset Blvd. • Hollywood, CA 90028 • (323)467-9375 Fax: (323)467-3962

DATE: CLIENT:
ARTIST: PRODUCER:
ENGINEER: 2nd ENGINEER:

TRACKS	SPEED	LEVEL	TAPE TYPE	NOISE RED	EMPHASIS	SAMPLE RATE	REEL NO.
TITLE				COMMENTS			

SMPTE START : : :
DOWNBEAT @ : : :
SMPTE END : : :
CODE TYPE 30 DF 25 24 NTSC
CODE SOURCE INT GEN EXT GEN TAPE
A1 A2
D1 D2

TRACK 1	TRACK 2	TRACK 3	TRACK 4	TRACK 5	TRACK 6	TRACK 7	TRACK 8
TRACK 9	TRACK 10	TRACK 11	TRACK 12	TRACK 13	TRACK 14	TRACK 15	TRACK 16
TRACK 17	TRACK 18	TRACK 19	TRACK 20	TRACK 21	TRACK 22	TRACK 23	TRACK 24
TRACK 25	TRACK 26	TRACK 27	TRACK 28	TRACK 29	TRACK 30	TRACK 31	TRACK 32
TRACK 33	TRACK 34	TRACK 35	TRACK 36	TRACK 37	TRACK 38	TRACK 39	TRACK 40
TRACK 41	TRACK 42	TRACK 43	TRACK 44	TRACK 45	TRACK 46	TRACK 47	TRACK 48

FIGURE 4.2
Example of a studio track log used for instrument/track assignments. Courtesy of Ocean Way Recording, www.oceanwayrecording.com.

At this point, the mic/line channels can be assigned to their respective tracks, making sure to fully document the assignments and other session info on the song or project's track sheet (Figure 4.2). If a DAW is to be used, the engineer should be sure to name each input track in the session for easy reference and track identification.

After all the assignments and labeling have been completed, the engineer then can begin the process of setting levels for each instrument and mic input by asking each musician to play solo by asking for a complete run-through of the song. Whilst these solo requests are taking place, be mindful that the engineer may wish to concentrate on the soloed instrument at hand, so consider refraining from practicing your latest riff. By placing each of the channel and master output faders to their unity (0 dB) setting and starting with the EQ settings at the flat position, the engineer can then check each of the track meter readings and adjust the mic preamp gains to their optimum gain while listening for preamp overload. If it's necessary, a gain pad can be inserted into the path in order to eliminate distortion.

After these levels have been set, a rough headphone mix can be made so that the musicians can hear themselves. Mic choice and/or placements can be changed or EQ settings can be adjusted (if absolutely necessary) to obtain the sound the producer wants on each instrument, and dynamic limiting or compression can be carefully inserted and adjusted for those channels that require dynamic attention. It's important to keep in mind that it's easier to change the dynamics of a track later during mixdown (particularly if the session is being recorded digitally) than to undo any changes that have been made during the recording phase.

Once this is done, the engineer and producer can listen for extraneous sounds (such as buzzes or hum from guitar amplifiers or squeaks from drum pedals) and eliminate them. Soloing the individual tracks can ease the process of selectively listening for such unwanted sounds, as well as for listening to the natural pickup of an instrument. If several mics are to be grouped into one or more tracks, the balance between them should be carefully set at this time.

After this procedure has been followed for all the instruments, the musicians should do a couple of practice "rundown" songs (or "Run Through" for us Brits! *Ed.*) so that the engineer and producer can listen to how the instruments sound together before being recorded. (If tape or disk space isn't a major concern, you might consider recording these tracks, as they might turn out to be your best takes—you just never know!) During the rundown, the engineer might consider soloing the various instruments and instrument combinations as a final check, and finally monitor all the instruments together. Careful changes in EQ can be made at this time (make sure to note these changes in the track sheet for future reference). These changes should be made sparingly, as final compensations are better made during the final mixdown phase.

While the song is being rundown, the engineer can make final adjustments to the recording levels and the headphone monitor mix. He/she can then check the headphone mix either by putting on a pair of headphones connected to the cue system or by routing the mix to the monitor loudspeakers. If the musicians can't hear themselves properly, the mix should be changed to satisfy their monitoring needs, regardless of their recorded levels. If several cue systems are available, multiple headphone mixes can be built up to satisfy those with different balance needs. During a loud session, the musicians might ask you to turn up their level (or the overall headphone level), so they can hear the mix above the ambient room leakage. It's important to note that high sound-pressure levels can cause the pitch of instruments to sound flat, so musicians might have trouble tuning or even singing with their headphones on. To avoid these problems, tuning shouldn't be done while listening through phones. The musicians should play their instruments at levels that they're accustomed to and adjust their headphone levels accordingly (many actually put only one cup over an ear, leaving the other ear free to hear the natural room sound).

The importance of proper headphone levels and a good cue balance can't be stressed enough, as they can either help or hinder a musician's overall performance. The same situation exists in the control room with respect to high monitor-speaker levels: Some instruments might sound out of tune, even when they aren't, and ear fatigue can easily impair your ability to properly judge sounds and relative balance.

During the practice rundown, it's also a good idea to ask the musician(s) to play through the entire song so the engineer will know where the breaks, bridges, and any point that's of particular importance might be. Making notes and even writing down or entering the timing numbers (possibly into a recorder's transport autolocator or into a DAW as session markers) can help speed up the process of finding a section during a take or overdub. The engineer can also pinpoint the loud sections at this time, so as to avoid any overloads. If compression or limiting is used, you might keep an ear open to ensure that the instruments don't trigger an undue amount of gain reduction (if the tracks are recorded digitally, gain reduction might be applied during mixdown). Even though an engineer might ask each musician to play as loudly as possible, they'll often play even louder when performing together. This fact may require further changes in the mic preamp gain, record level, and compression/limiting threshold. Soloing each mic and listening for leakage can help to check for separation and leakage between the instruments. If necessary, the relative positions of mics, instruments, and baffles can be changed at this time.

Electric and Electronic Instruments

Electric instruments (such as guitars and bass guitars) generally have mid-level, unbalanced, high-impedance outputs that can be directly recorded in the studio without using their amplifiers, via a direct box (DI). As we've already learned, these devices convert line-level, high-impedance output signals into a low-impedance, balanced signal that can be fed directly into a console's mic preamp or external preamp. Instruments often are recorded "direct" in order to avoid instrument leakage problems in the studio, to reduce noise and distortion that can occur when miking an amp, or simply to get a "clean and tight" direct sound.

Electronic instruments (such as synths, samplers, and effects boxes) more closely match the impedance and levels of studio equipment and can be plugged directly into a console or interface line-level input. Within the studio itself, DI boxes are generally still considered to be the best way to insert a signal into the console.

Any of these instrument types can be played in the control room while listening over the main studio monitor speakers without fear of leakage. If you prefer, the signal of a DI can be split into two paths: one that can be directly inserted into the console/interface and another that can be fed to an instrument speaker in the studio (Figure 4.3). This technique makes it possible for the both direct and miked pickup to be combined and blended, giving you the benefits of a clean, direct sound with the added rougher, gutsier sound of a miked amp.

On most guitars, the best tone and lowest hum pickup for a direct connection occurs when the instrument volume control is fully turned up. Because guitar tone controls often use a variable treble roll-off, leaving the tone controls at the treble setting and using a combination of console EQ and different guitar pickups to vary the tone will often yield the maximum amount of control over the sound. If the treble is rolled off at the guitar, boosting the highs with EQ will often increase the pickup noise.

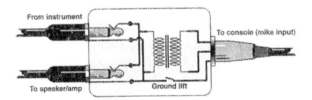

FIGURE 4.3
Basic schematic for a direct box (DI).

©2009 David Miles Huber

Drums

During the past several decades, drums have undergone a substantial change with regard to playing technique, miking technique, and choice of acoustic recording environment. In the 1960s and 1970s the drum set was placed in a small isolation room called a drum booth. This booth acoustically isolated the instrument from the rest of the studio and had the effect of tightening the drum sound because of the limited space (and often dead acoustics). The drum booth also physically isolated the musician from the studio, which often caused the musician to feel removed and less involved in the action. Today, many engineers and producers have moved the drum set out of smaller isolation (iso) booths and back into larger open studio areas where the sound can fully develop and combine with the studio's own acoustics. In many cases, this effect can be exaggerated by placing a distant mic pair in the room (a technique that often produces a fuller, larger-than-life sound, especially in surround sound).

Before a session begins, the drummer should tune each drum while the mics and baffles for the other instruments are being set up. Each drumhead should be adjusted for the desired pitch and for constant tension around the rim by hitting the head at various points around its edge and adjusting the lugs for the same pitch all around the head. Once the drums are tuned, the engineer should listen to each drum individually to make sure that there are no buzzes, rattles, or resonant after rings. Drums that sound great in live performance may not sound nearly as good when being close miked. In a live performance, the rattles and rings are covered up by the other instruments and are lost before the sound reaches the listener. Close miking, on the other hand, picks up the noises as well as the desired sound.

If tuning the drums doesn't bring the extraneous noises or rings under control, duct or masking tape can be used to dampen them. Pieces of cloth, dampening rings, paper towels, or a wallet can also be taped to a head in various locations (which is determined by experimentation) to eliminate rings and buzzes. Although head damping has been used extensively in the past, present methods use this damping technique more discreetly and will often combine damping with proper design and tuning styles (all of which are the artist's personal call).

During a session, it's best to remove the damping mechanisms that are built into most drum sets, as they apply tension to only one spot on the head and unbalance its tension. These built-in dampers often vibrate when the head is hit and are a chief source of rattles. Removing the front head and placing a blanket or other damping material inside the drum (so that it's pressing against the head) can often dampen the kick drum. Adjusting the amount of material can vary the sound from being a resonant boom to a thick, dull thud. Kick drums are usually (but not always) recorded with their front heads removed, while other drums are recorded with their bottom heads either on or off. Tuning the drums is more difficult if two heads are used because the head tensions often interact; however, they will often produce a more resonant tone. After the drums have been tuned, the mikes can be put into position. It's important to keep the mics out of the drummer's way, or they might be hit by a stick or moved out of position during the performance.

The Final Check

The placement of musicians and instruments will often vary from one studio and/ or session to the next because of the number of instruments, isolation (or lack thereof) among instruments, and the degree of visual contact that's needed. If additional isolation (beyond careful microphone placement) is needed, flats and baffles can be placed between instruments in order to prevent loud sound sources from spilling over into other open mikes. Alternatively, the instrument or instruments could be placed into separate iso-rooms and/or booths, or they could be overdubbed at a later time.

During a session that involves several musicians, the setup should allow them to see and interact with each other as much as possible. It's extremely important that they will be able to give and receive visual cues and "feel the vibe." The instrument/mic placement, baffle arrangement, and possibly room acoustics (which can often be modified by placing absorbers in the room) will depend on personal preferences, as well as on the type of sound the producer wants. For example:

- If the mics are close to the instrument and the baffles are packed in close, a tight sound with good separation will often be achieved.
- A looser, more "live" sound (along with an increase in leakage) will occur when the mics and baffles are placed farther away.
- An especially loud instrument can be isolated by putting it in an unused iso-room or vocal or instrument booth. Electronic amps that are played at high volumes can also be recorded in such a room. Alternatively, the amps and mics can be surrounded on several sides by sound baffles and (if needed) a top can be put on the "box."
- An amp and the mic can be covered with a blanket or other flexible sound-absorbing material (Figure 4.4), so that there's a clear path between the amplifier and the mic.
- Separation can be achieved by placing the softer instruments in an iso-room or by plugging otherwise loud electronic instruments directly into the console via a D.I. (direct injection) box, thereby bypassing the miked amp.
- Piano leakage can be similarly reduced by placing one or more mics inside it, putting the lid on its short support stick and covering it with blankets (Figure 4.5), or by placing it in an iso-room.

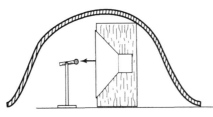

FIGURE 4.4
Isolating an instrument amplifier by covering it with a sound-absorbing blanket.

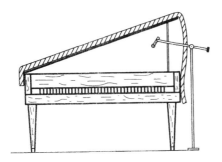

FIGURE 4.5
Preventing leakage from entering into a piano mic.

Obviously, these examples can only suggest the number of possibilities that can occur during a session. For example, you might collectively choose not to isolate the instruments, and instead place the instruments in an acoustically "live" room. This approach will require that you carefully place the mics in order to control leakage; however, the result will often yield a live and present sound. As artists, the choices belong to you, the producer, the group, and (possibly) the production/distribution company.

From a technical/engineering standpoint, if analog tape machines are to be used, they should be cleaned, demagnetized, and (if necessary) aligned for the specific tape formulation that's to be used for the session (it's preferable to use the actual session tape itself). Generally it's a good idea to record a 100-Hz, 1-kHz, and 10-kHz tone at 0 VU on all tracks at the beginning of the tape to indicate the proper operating level. If it's necessary to overdub or mix at another studio, the tones will make it easier for the engineer to calibrate the unknown tape machine to your reference tones. SMPTE time code should also be stripped before the session starts (making sure to know the proper TC rate for the session).

If tape-based digital recorders are to be used, make sure that the tapes have been properly formatted well before the session starts. Although it's true that certain recorders can record onto unformatted tapes in a single pass or in a special update mode, it's never as easy as having tapes that are preformatted.

If a hard-disk DAW is to be used, make sure that the proper sample and bit rates are chosen for the session (this should be discussed beforehand with the group and/or producer). If an external drive (or drives) has been brought in for the session, make sure that it has been formatted and is ready to roll. Remember, the idea is to work out the kinks and to simplify technology as much as possible. (Murphy's Law is and always will be alive and well in any production facility).

RECORDING

It's obviously a foregone conclusion that no two recording sessions will ever be exactly alike. In fact, in keeping with the "no rules" rule, they're often radically different from each other. During the *recording session*, the engineer watches the level indicators and (only if necessary) controls the faders to keep from overloading the media. It's also an engineer's job to act as another set of production ears by listening both for performance and quality factors. If the producer doesn't notice a mistake in the performance, the engineer just might catch it and point it out. The engineer should try to be helpful and remember that the producer and/or the band will have the final say, and that their final judgment of the quality of a performance or recording must be accepted.

Once recording is under way, at the beginning of each performance, the name of the song and a take number are recorded to tape for easy identification (a process that's often referred to as *slating* the tape). A take sheet should be carefully kept to note the position of the take on a tape. Comments are also written onto this sheet to describe the producer's opinion of the performance, as well as other information of importance. In fact, the concept of documentation can't be overstressed. The diligent documentation of a session might not be one of the most fun things to do, but should changes be needed in the future (knowing which effect was used on a vocal, how an instrument was placed in a surround field, what mic or pickup combo was used to capture that killer guitar riff, etc.), would come in mighty handy.

From the musician's standpoint, here are a few additional pointers that can help the session go more smoothly:

- It's always best to get the right sound and vibe onto tape or disk during the session. If you need to do another take, do it! If you need to change a mic, change it. "Fixing in the mix" will usually apply a band-aid to an ailing track.
- Know when to quit! If you're pushing too hard or are tired, it'll often show.
- Technology doesn't always make a good track; feeling, emotion, and musicality does.
- Beware of adding new parts or tracks onto a piece that's "just right." Remember, too much is too much!
- Leave plenty of time for the vocal track(s). It's not uncommon for a group to spend most of their time and budget on getting the perfect drum or guitar sound. It takes time and a clear focus to get the vocals right.
- If you mess up on a part, keep going, you might be able to fix the bad part by punching-in. If it's really that bad, the engineer or producer will stop you.

In his *EQ Magazine* article "The performance curve: How do you know which take is the one?," Craig Anderton laid out his experiences on how different musicians will deal with the process of delivering a performance over time. Being in front of a mic isn't always easy and we all deal with it differently. Here's a basic outline of his findings:

- *Curves up ahead*: With this type of performer, the first couple of takes are pretty good, then start to go downhill before ramping back up again, until they hit their peak before going downhill really fast.
- *The quick starter*: This type starts out strong and doesn't improve over time in later performances. Live performers often fall into this category, as they're conditioned to get it right the first time.
- *The long ramp-up*: These musicians often take awhile to warm up to a performance. After they hit their stride, you might get a killer take or a few great ones that can be composited, or comp'd together into the perfect take.
- *Anything goes*: This category can vary widely within a performance. Often snippets can be taken from several takes into a single composite. You want to record everything with this type of performer, because you just never know what gem (or bad take) you'll end up with.
- *Rock steady*: This one represents the consummate pro that's fully practiced and delivers a performance that doesn't waver from one take to the next; however, you might record several takes to see which one has the most feeling.

From the above examples, we can quickly draw the obvious conclusion that there are all types of performers, and that it takes a qualified and experienced producer and/or engineer to intuit just which type is in front of the mic and to draw the best possible performance from him/her.

OVERDUBBING

Overdubbing (Figure 4.6) is used to add more instruments to a performance after the basic tracks have been recorded. These additional tracks are added by monitoring the previously recorded tape tracks (usually over headphones) while simultaneously recording new, doubled, or augmented instruments and/or vocals onto one or more available tracks.

In an overdub (OD) session, the same procedure is followed for mic selection, placement, EQ, and level as they would occur during a recording session. If only one instrument is to be overdubbed, the problem of having other instrument tracks leak into the new track won't exist. However, leakage can occur if the musician's headphones are too loud or aren't seated properly on his or her head.

The natural ambience of the session should be taken into account during an overdub. If the original tracks are made from a natural, roomy ensemble, it could be distracting to hear an added track that was obviously laid down in a different (usually deader) room environment.

If the recorder to be used is analog, it should be placed in the master sync mode (thereby reproducing the previously recorded tracks from the record head in sync). The master sync mode is set either at the recorder or using its autolocator/remote control. Usually, the tape machine can automatically switch between monitoring the source (signals being fed to the recorder or console) and tape/sync (signals coming from the playback or record/sync heads). The control room monitor mix should prominently feature the instrument(s) that's being recorded, so mistakes can be easily heard. During the initial rundown, the headphone cue mix can be adjusted to fit the musician's personal taste.

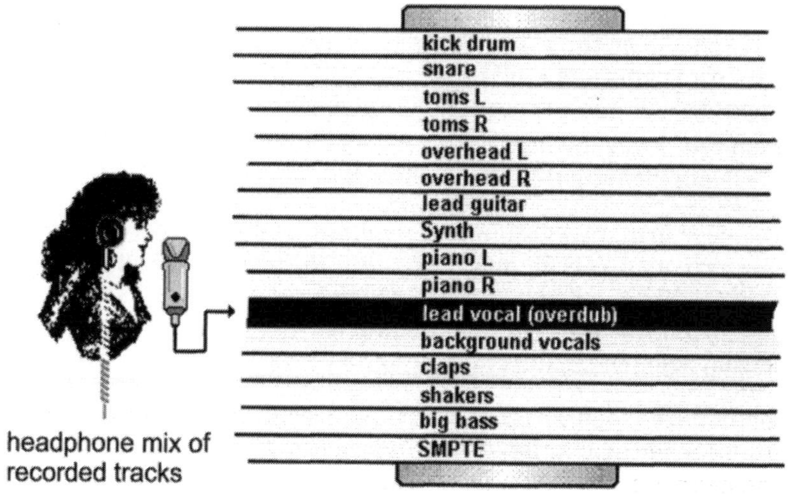

kick drum
snare
toms L
toms R
overhead L
overhead R
lead guitar
Synth
piano L
piano R
lead vocal (overdub)
background vocals
claps
shakers
big bass
SMPTE

headphone mix of
recorded tracks

FIGURE 4.6
Overdubbing allows instruments to be added to existing tracks on a multitrack recording medium.

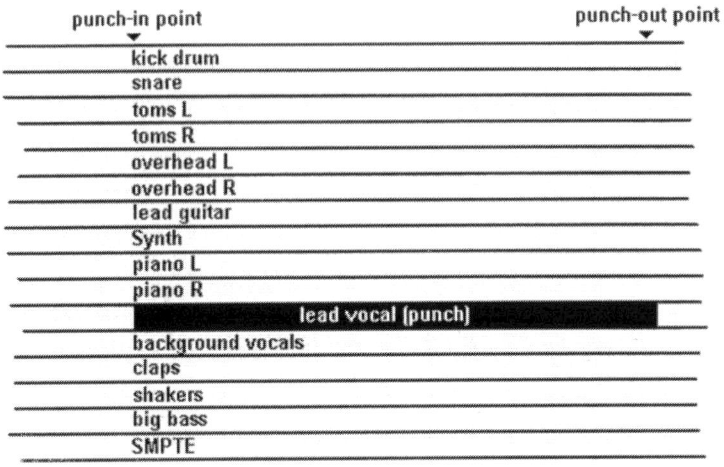

FIGURE 4.7
Punch-ins let you selectively replace material and correct mistakes.

Punching-In

Should a mistake or bad take be recorded onto the new track, it's a simple matter to start over and re-record over the unwanted take. If only a small part of the take was bad, it's easy to *punch-in* (Figure 4.7), a process that can:

- Silently enter the record mode at a pre-determined point
- Record over the unwanted portion of the take
- Silently fall back out of record at a pre-determined point.

A punch can be manually performed on most recording systems; however, newer tape machines and DAW systems can be programmed to automatically go into and fall out of record at a pre-determined time.

When punching-in, any number of variables can come into play. If the instrument is an overdub, it's often easy to punch the track without fear of any consequences. If an offending instrument is part of a group or ensemble, leakage from the original instrument could find its way into adjacent tracks, making a punch difficult or impossible. In such a situation, it's usually best to re-record the piece, pick up at a point just before the bad section and splice or insert it into the original recording, or attempt to punch the section using the entire ensemble.

It's often best to punch-in on a section immediately after the take has been recorded, as changes in mic choice, mic position, or the general mood of the session can lead to a bad punch that will be hard to correct. If this isn't possible, make sure that the engineer or producer carefully documents the mic choice, placement, preamp type, etc., before moving onto the next section: You'll be glad you did.

Performing a "punch" on a tape-based recorder should always be done with care. Allowing the tape to record too long a passage could possibly cut off a section of the following acceptable track and require that the following section be likewise overdubbed. Stopping it short could cut off the natural reverb trail of the final

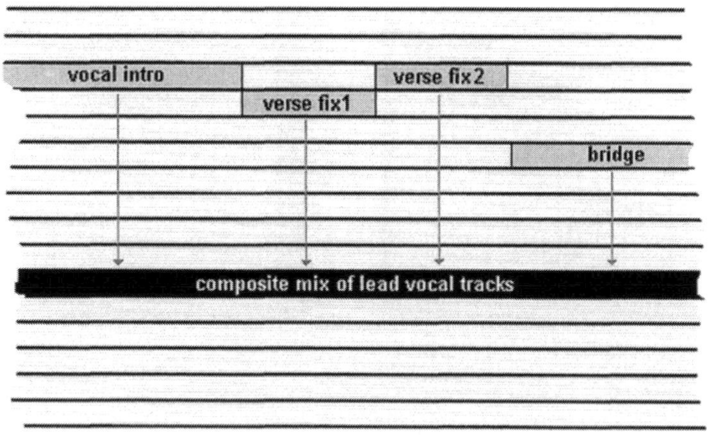

FIGURE 4.8
A single composite track can be created from several partially acceptable takes.

note. Performing a punch using a DAW is often far easier, as the leading and trailing edge of the punch can often be manually adjusted to expose or hide sections after the punch has been performed. Most modern recording media can also be programmed to perform the punch under automated control. This built-in repeatability can go a long way toward reducing operator error and its associated tension.

An additional technique that can be used in conjunction with (or as an alternative to) punching-in over a recorded track is to overdub the instrument onto another available track or set of tracks. The advantage of recording onto another track (if any are available) is that a good or marginal take can be saved and the musician can try to improve the performance, rather than having to erase the previous take in order to improve it. When several tracks of the overdub have been saved, different sections of each take might be better than others. These can be combined into a single, composite (comp) take by playing the tracks back in the sync mode, mixing and muting them together at the console (often with the help of automation), and recording them on another tape track (Figure 4.8). It should also be noted that the job of recording multiple takes and then combining them into a single composite is also well suited to a DAW, which can often record unlimited takes (without using up physical tape tracks) and then combine them into a composite track with relative ease. (*Note*: It's often a good idea to export the comp takes into a single, composite track that can be easily archived and imported from one DAW platform to another.)

Bouncing Tracks

Another procedure that's an extension of the composite process (when using a tape-based system) is the concept of track *bouncing* or *ping-ponging*. The process of bouncing tracks is often used to mix a group or an entire set of basic tracks down to one track, a stereo track pair, or grouping of several tracks, usually when the number of available tape tracks are limited.

Bouncing can be performed either to make the final mixdown easier (by grouping instruments together onto one or more tracks, as seen in Figure 4.9) or to open up

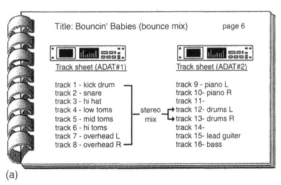
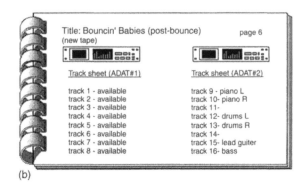

(a)   (b)

FIGURE 4.9
Track bouncing is a common production technique that's used in tape-based multitrack recording.
(a) Instruments can be grouped together onto one or more tracks or (b) bouncing can be used to expand
the number of available tracks and thus allow for additional overdubs.

needed tape tracks by bouncing similar instrument groups down to one or more tracks. This frees up the originally recorded tape tracks for overdubs.

For example, suppose that we have two Modular Digital Multitrack (MDM) digital recorders with which to do a demo recording (giving us a total of 16-tracks). During the basic track session, let's say that we decided to record the drums onto all 8-tracks of MDM #1 and then record a stereo piano, bass, and lead guitar (4-tracks) onto separate tracks of MDM #2. After these 12-tracks have been recorded, we could then go back and mix the drums down to a stereo submix pair of the 4 available tracks on MDM #2. Once this is done, the original drum tape can be set aside. (Whenever possible it's always a good idea to keep the original recordings intact, just in case you need them in the future or in case of an accidental punch error—Murphy's Law, you know!) Once we've put a new tape into MDM #1, we can go about the task of recording onto 10-tracks (8 newly opened tracks of MDM #1 and 2-tracks on #2, as shown in Figure 4.9).

When using a DAW, bouncing isn't usually necessary as there are large numbers of virtual tracks that can be used. In this situation, any number of tracks can be corralled together into groups that can be easily controlled in level from a single group fader.

MIXDOWN

After all the tracks for a song have been recorded, the DAW, multitrack tape or other medium can be mixed down to a rough mix perhaps for you to listen to, unless the mixing is to be done in the same session by the same engineer. In which case the engineer will make a mix to mono, stereo, or surround for subsequent duplication and distribution to the consumer.

However, it might have been decided that a specialist mix engineer should be employed to enhance the status of the track yet further (specialist mix engineers can carry weight in A&R circles and it is believed their mixes will translate exceptionally well to radio and increase exposure and ultimately sales). Roey Izhaki addresses this type of engineer in Chapter 5B.

When beginning a mixdown using an analog multitrack and/or mastering recorder, it's customary to demagnetize, clean, and align the machine. Once this is done, the engineer should record 0-VU level tones at the head of the mixdown reel at 1 kHz, 10 kHz, and 100 Hz. This makes it possible for the mastering engineer to align the tape playback machine at these reference frequencies, resulting in a proper playback EQ and levels.

Once the engineer is ready, the console can be placed into the mixdown mode (or each input module can be switched to the line or tape position) and the fader label strips can be labeled with their respective instrument names. Channel and group faders should be set to unity gain (0 dB) and the master output faders should likewise be set with the monitor section being switched to feed the mixdown signal to the appropriate speakers. If a DAW is being used to create the final mix, a basic array of effects can be programmed into the session mixer and, if available, a controller surface can be used to facilitate the mix by giving you hands-on control.

The engineer can then set up a rough mix of the song by adjusting the levels and the spatial pan positions. The producer then listens to this mix and might ask the engineer to make specific changes. The instruments are often soloed one by one or in groups, allowing any necessary EQ changes to be made. The engineer, producer, and possibly the group can then begin the cooperative process of "building" the mix into its final form. Compression and limiting can be used on individual instruments as required (either to make them sound fuller, more consistent in level, or to prevent them from overloading the mix when raised to the desired level). At this point, the console's automation features can be used (if available). Once the mix begins to take shape, reverb and other effects types can be added to shape and add ambience in order to give close-miked sounds a more "live," spacious feeling, as well as to help blend the instruments.

During this rough mix process, it is advisable to leave the engineer to it for a number of hours. Whilst mixing is an exciting process, it is also both time consuming and engrossing for the engineer. Focus is important and it is both difficult and distracting to mix with the band chatting on the sofa. It is best for the band to return to the studio for a listen at points throughout the mix. It is an occasion to see how the mix is progressing, and to offer an objective ear to the process having been away from constantly working on it.

If the mix isn't automation assisted, the fader settings will have to be changed during the mix in real time. This means that the engineer will have to memorize the various fader moves (often noting the transport counter to keep track of transition times). If more changes are needed than the engineer can handle alone, the assistant, producer, or artist (who probably knows the transition times better than anyone) can help by controlling certain faders or letting you know when a transition is coming up.

It's usually best, however, if the producer is given as few tasks as possible, so that he/she can concentrate fully on the music rather than the physical mechanics of the mix. The engineer then listens to the mix from a technical standpoint to detect any sounds or noises that shouldn't be present. If noises are recorded on tracks that aren't used during a section of a song, these tracks can be muted until needed. After the engineer practices the song enough to determine and learn all the changes, the mix can be recorded and faded at the end. The engineer might not want to fade the song during mixdown, as it will usually be performed after being transferred to a DAW (which can perform a fade much more smoothly than even the smoothest hand). Of course, if automation is available or if the mix is performed using a DAW, all of these moves can be performed much more easily and with full repeatability.

It's usually important that levels should be as consistent as possible between the various takes and songs, and it's often wise to monitor at consistent, moderate listening levels. This is due to the variations in our ear's frequency response at different sound-pressure levels, which could result in inconsistencies between song balances. Ideally, the control room monitor level should be the same as might be heard at home, over the radio, or in the car (between 70 and 90 dB SPL), although certain music styles will "want" to be listened to at higher levels. Once the final mix or completed project master is made, you might want to listen to it over different speaker systems (ranging from the smallest to the biggest or baddest you can find). It's usually a good idea to run off a few copies for the producer and the band members to listen to at home and in their cars. In addition, the mix should be tested for mono–stereo/surround compatibility (when using any mix format) to see what changes in instrumental balances might have occurred. If there are any changes in frequency balances or if phase becomes a problem when the mix is played in mono, the original mix might have to be modified.

SEQUENCE EDITING

After all the final mixes for a recording have been completed, a final edited master could be assembled. The producer and artists begin the process by listening to the songs and deciding on a final sequence, based on their tempos, musical keys, how they flow into one another, and which songs will best attract the listener's attention. Once this is agreed upon, the process of assembling the final master can begin whether that is in-house or at a dedicating mastering facility.

MASTERING

Although the subject of mastering is covered in-depth within Chapters 6 & 7, the fact is that the producer and/or artist will be faced with the question of whether to hire the services of a qualified or well-known mastering engineer to put the finishing touches on a master recording, or to make use of the talents at hand (i.e., producer, engineer, and artist) to finesse the various songs into a final product statement. These questions should be thoroughly discussed in the pre-planning phase, allowing for an on-the-spot change of plans.

A FINAL WORD ON PROFESSIONALISM

Before we close this chapter, there's one more subject that I'd like to touch upon—perhaps the most important one of all—*professional demeanor*. Without a doubt, the life and job of a typical engineer, producer, or musician isn't always an easy one. It often involves long hours and extended concentration with people who, more often than not, are new acquaintances. In short, it can be a high-pressure job. On the flip side, it's one that's often full of new experiences, with demands changing on almost a daily basis, and often involves you with exciting people who feel passionately about their art and chosen profession.

It's been my observation (and that of many I've known) that the best qualities that can be exhibited by anyone in "The Biz" are:

- Having an innate willingness to experiment
- Being open to new ideas (flexibility)
- Having a sense of humor
- Having an even temperament (this often translates as patience)
- Willing to communicate with others
- Being able to convey and understand the basic nuances of people from all walks of life and with many different temperaments.

The best advice I can possibly give is to *be open, be patient, and above all, be yourself.* Also, be extra patient with yourself. If you don't know something, ask. If you made a mistake (trust me, you will, we all do) admit it and don't be hard on yourself. It's all part of the process of learning and gaining experience. This last piece of advice might not be as popular as the others, but it may come in handy someday.

FURTHER READING

This edited chapter is from Huber, D. & Runstein, R. (2005). *Modern Recording Techniques*. Focal Press, Oxford.
Please also refer to www.modrec.com for more information.

Chapter 5

A. The Mix Session: What's Going On?

Roey Izhaki

In This Chapter

In this chapter, Roey Izhaki starts by discussing how engineers learn to mix. In the second part, Roey introduces some of the procedures and creative decision making, which makes mixing a unique and rich part of the production process. With this knowledge on board, you will begin to understand the process and what to listen for when your work is at the faders.

INTRODUCTION

An analogy can be made between the process of learning a new language and that of learning to mix. At the beginning, the engineer will start with no or very little knowledge and nothing seems to make sense.

With language, you cannot understand a single sentence or even separate the words within it; just like if you play a mix to most people they will not be able to hear the reverb or compression (they've hardly ever focused on these sonic aspects before, that's if they've heard of reverbs or compressors). After learning some individual words and when to use them, you find yourself able to identify them in a sentence on the same basis, you start learning how to use compressors and reverbs, and you start hearing these in mixes.

To pronounce a new word can be hard, since it is not easy to notice the subtle pronunciation differences of a new language, but after hearing and repeating a word for 20 times, you get it right; likewise, after compressing 20 vocal tracks, you start hearing degrees of compression and you can tell which compression suits the best. Then, you learn grammar which enables you to connect all the words together and construct a coherent sentence; this highlights the point when all your mixing techniques help you to craft a mix as a whole.

Finally, since in a conversation there is more than one sentence, the richer your vocabulary is and the stronger your grammar, the more sentences you are able to construct properly. In mixing, the more techniques and tools you learn and the more mixes you craft, the better your mixes become. All in all, the more you learn and practice a new language, the better you become at it. The same is for mixing.

WHAT MAKES A GREAT MIXING ENGINEER?

World-class mixing engineers might charge for a single album twice the yearly minimum wage in their respective country. Some mixing engineers also ask for points—a percentage from album sale revenue. Across both sides of the Atlantic, a revered mixing engineer can make a yearly figure of six digits. These people are not being paid for nothing—the amount of knowledge, experience and talent they have is immense. Record labels reward them for that, and in exchange see higher sales. As mixing is a separate stage in the production chain, it is clear why a specialized person might do it. Yet, mixing is such a huge area that there is no wonder why some people choose to devote themselves solely to it—the amount of *knowledge* and *practice* required to make a great mixing engineer is enough to fill a whole life span.

Primarily, the creative part of mixing revolves around three steps—vision, evaluation and action—as shown in Figure 5.1. The ability to successfully go through these steps can lead to an outstanding mix. But a great mixing engineer will need a notch more than that, especially if hired. These steps are explained, along with the requisite qualities that make a great mixing engineer, in the following sections.

Mixing Vision

As mentioned earlier in this book, composing can entail different approaches. One of them involves utilizing an instrument, say a piano, then either by means of trial

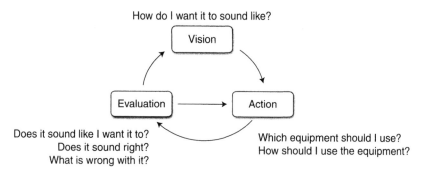

FIGURE 5.1
The three steps around which the creative part of mixing revolves.

and error or based on some music theory, coming up with a chord structure and melody lines. Another approach involves imagining or thinking of a specific chord or melody and only then playing it (or in the case of trained composers—writing it straight to paper). Many composers and songwriters gave the latter account on the process of composition—first imagine, then play or write.

We can make an analogy between these two different composing methods and mixing. If we take the equalization process of a snare, for example, the first approach involves sweeping through the frequencies, then choosing whatever frequency seems most appealing. The second approach involves first imagining the desired sound and only then approaching the EQ in order to attain it. Put another way, the first approach might entail thinking such as "OK, let's try to boost on this frequency and see what happens," while the other might entail "I can truly imagine how I want this snare to sound like—it should have less body, but sound more crispy." Just like composers can imagine the music before it is played, a mixing engineer can imagine sounds before taking any action—a big part of mixing vision. Mixing vision is primarily concerned with the fundamental question: how do I want it to sound like? The answers could be many—soft, powerful, clean or intense are just a few examples. But mixing vision cannot be defined by words alone—it is a sonic imagination, which later crystallizes through the process of mixing. The mixing vision is a bipartisan affair. Both the artist and the production team must share in their vision for the final mix and continue dialog throughout.

The selection of ways we have to alter sounds is great—equalizing, compressing, gating, distorting, adding reverb or chorus and many more. So which type of treatment should the engineer pick? There are also infinite ways (in the analog domain anyway) within each category—the frequency, gain and Q controls on a parametric equalizer provide millions of possible combinations. So why should we choose a specific combination and not another? Surely, equalizing something in a way that makes it sound right does not assure that different equalization would not make it sound better. Mixing vision gives the answer to these questions, "because this is how we imagined it, this is how we wanted it to sound like."

One of the shortcomings novice engineers might have is lack of imagination. The process of mixing for them is a trial-and-error affair between acting and evaluating

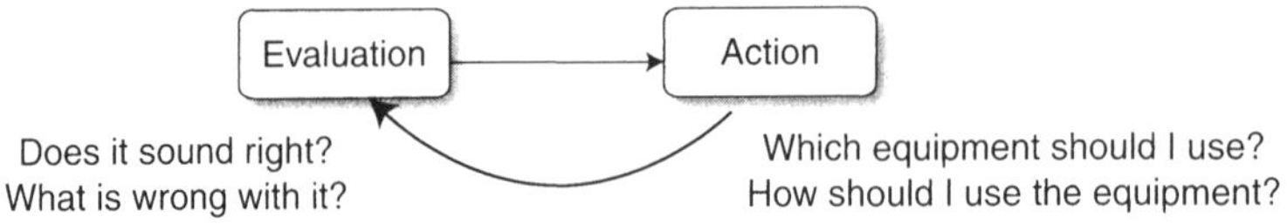

FIGURE 5.2
The novice approach to the creative part of mixing might be missing in a mixing vision, therefore it only involves two stages.

(Figure 5.2). But how can one critically evaluate something without a clear idea of what one wants? Having no mixing vision can make mixing a very frustrating hit-and-miss process.

The Skills to Evaluate Sounds

The ability to help craft a good mix is based on countless evaluations. One basic question, often asked at the beginning of the mixing process, is "What's wrong with it?". A possible answer could be "the highs on the cymbals are harsh" or "the frequency spectrum of the mix is too heavy on the mids." From the endless amount of treatment possibilities we have, focusing on rectifying the wrongs provides a good starting point. It can also prevent the novice from doing things for no good reason or with no specific aim—for example, equalizing something that does not really require equalization.

At points it might be hard to tell what is wrong with the mix, in which case our mixing vision provides the basis for our actions. After applying a specific treatment, the novice might ask "Does it sound right?", whilst the veteran might also ask "Does it sound the way I want it to?" Clearly, the veteran has an advantage here since the latter question is less abstract.

Mastering His or Her Tools and Knowledge of Other Common Tools

With or without a clear mixing vision, we perform many actions in order to alter existing sounds. When choosing a reverb for vocals, the novice might go through all the available presets on a reverb emulator. These can easily exceed the count of 50, and the whole process takes some time. The veteran, on the other hand, will probably quickly access a specific emulator and choose a familiar preset, a bit of tweaking and the task is done. It takes very little time. Experienced mixing engineers know, or can very quickly conclude, which tool would do the best job in a specific situation; they can quickly answer the question: "Which equipment should I use?"

Nevertheless, professional mixing engineers do not always work in their native environment as they sometimes travel to work in different studios. Although at times they take their favorite gear with them, a big part of the mix is done using in-house equipment. Professional mixing engineers therefore have to be familiar with common tools found in commercial environment.

Mastering one's tools does not only stand for the ability to pick the right tool for a specific task, but also the expertise to employ the equipment in the best way ("How should I use the equipment?"). Whether to choose high-shelving or high-pass characteristic on an equalizer is one example. The experience to know that a specific compressor can work well on drums when more than one ratio button is pressed is another example.

It is also worth discussing the quantity of tools at our disposal. Nowadays, Digital Audio Workstation (DAW) users seem to have much more selection than those mixing using hardware. Not only plug-ins are cheaper, but a plug-in can have as many instances the computer can handle. Once a specific hardware processor is used for a specific track on the desk, it cannot be used simultaneously on a different track. While in an analog studio a mixing engineer might have around three favorite compressors to choose from when processing vocals, DAW users might have ten. Learning each of these compressors—*understanding* each of them—takes time; just reading the manual of some tools can take a whole day. Having an extensive amount of tools can mean that none of them are being used to their best extent because there is no time to learn and properly experiment with them all. Mixing is a simple process that only requires a pair of trained ears and a few quality tools. When it comes to the quantity of tools, more can be less and less can be more.

Theoretical Knowledge

The main four questions are:

1. When clipping shows on the master track in an audio sequencer, is it the master fader or all of the channel faders that should be brought down?
2. For more realistic results, should one or many reverb emulators be used?
3. Why and when stereo linking should be engaged on a compressor?
4. When should dither be applied?

To say that every single mixing engineer knows the answers to these questions would be a lie, but most do. The same is for saying that you cannot craft an outstanding mix without this knowledge; in our mixing community there are more than a few highly successful engineers who do not know the answers to many theoretical questions. But it would also be a lie to say that knowing the answers to these questions would not be an advantage. Like talent, knowledge is always a blessing. In this competitive field, knowledge is sometimes what makes the difference between two equally talented engineers. Some acquire knowledge through condensed educational programs, others learn little-by-little as they mix. But all mixing engineers are compulsive learners; if the ratio on a compressor is set to 1:1, one would spend hours trying to figure out why no other control has effect. Surely knowing the difference between shelving and pass filter is a handy one. And dither does affect the final mix quality. It would seem unreasonable for a mastering engineer not to know when to apply dither; mixing engineers should know it, especially as they are likely to apply it too, across more tracks.

Interpersonal Skills

Studio producers need an enormous capacity to deal and interact with many people who are known to have different abilities, moods and degrees of dedication. Mixing engineers, on the contrary, tend to work on their own and only on occasion mix in front of the client, whether that be the artist, A&R, producer or all three. Like any job that involves interaction with people, mixing also requires good interpersonal skills.

When the band comes to listen to the mix, it should not come as a surprise if each of the musicians think their instrument is not loud enough (after all, they are used to their instrument being the loudest whether on stage or through the cans in the live room). Even the old tricks of limiting the mix or blasting the full-range speakers do not always work to appease. In more than a few occasions, artists and A&R remark on the work of mixing engineers with the same rational of accountants commenting on the work of the graphic designers whom they hired. While the feedback from fresh ears can sometimes be surprisingly constructive, at other times the mixing engineer is safe in the personal knowledge that sometimes the client's comments are either technically or artistically inappropriate. As such things can easily become personal—mixing engineers, like artists, can become extremely attached to their work. Interpersonal skills can solve artistic disagreements, and help some engineers to demonstrate their views better. But if the artist does not like some aspect of the mix, he or she will have to listen to this disliked aspect for the rest of their life unless the mixing engineer compromises. The client-always-right law works in mixing all the same—after all, a displeased client is a lost client. However, it is always worth trusting the ability of the mixing engineer, as their experience will tell whether your bass guitar is loud enough or not. Listening to the mix in different environments later will allow you to truly assess whether you're right or not.

The Ability to Work Fast

The beginning stages can be hard when learning something new—most guitar players experienced some frustration before they could change chords quickly enough or produce a clean sound. It is not fun to work on a single verse for a whole day and still be unhappy with the mix at the end. As mixing experience accumulates, it takes less time to choose tools and utilize them to achieve the desired sound. Also, the engineer's mixing vision becomes sharper and we can crystallize it quickly. Altogether, each task takes less time, which leaves more time to elevate the mix or experiment. Needless to say, the ability to work fast is essential for hired mixing engineers, who work under busy schedules and strict deadlines.

METHODS OF LEARNING MIXING

Reading About Mixing

Audio literature is great. Books, magazine articles, even Internet forums can sometimes include extremely valuable theory, concepts, ideas or tips. Reading about mixing will not make a great mixing engineer, in the same way reading about cookery will not make a great cook. Reading about mixing gives us a better chance to understand core concepts and how the engineer operates the tools, but the one thing it does not do is improve our sonic skills. Some reading in mixing will allow you to understand the process unfolding in front of you.

Reading manuals is also an important practice, although unfortunately many engineers and musicians alike choose to neglect it. The basic aim of a manual is to teach us how to use equipment, and sometimes also how to use it right or more appropriately. In their manuals, many companies present some practical advice, whether on their products or on mixing in general. Sometimes the controls of a certain tool are not straightforward and it can take eternity to understand what their function is.

Reading and Hearing

This book is an example of this method. An aural demonstration of mixing-related issues provides a chance to develop critical evaluation skills and better understanding of sonic concepts. While this method can contribute to all stages of mixing—vision, action and evaluation—it is a passive way of learning since it does not involve active mixing.

Seeing and Hearing

Watching other people mix is another way to learn. Many young people wish to work in a studio so they can learn from the experience. Listening to others while they craft their mix is great, but it comes with two cautions. First, it is impossible to enter people's minds—while watching them mixing it might be possible to understand what they are doing, but not why they do it. Mixing vision and experience are nontransferable. Second, if we take into account the tricks and tips already published, what is left to learn from these experienced people is mostly their own unique techniques rather than mixing as a whole. True, learning the secret techniques of top-of-the-line mixing engineers is great, but only if these are used in the right context later. There is some belief that the greatest mixing engineers craft amazing mixes because of secret techniques. In practice, these amazing mixes are the outcome of extensive understanding of, and experience in basic techniques. Most of the individual's secret techniques often only add some degree of polish and a distinguished sonic stamp.

MIXING ANALYSIS

Earlier in this book, we spoke of listening analysis. Here we expand upon this with specific attention to mixing.

Sometimes learning the techniques of an art makes it hard to look at the art as a whole. For example, while watching a movie, film students may analyze camera movements, lighting, edits, lip sync or acting. It can be hard for those students to stop analyzing and just enjoy movies like when they were fascinated kids. However, many mixing engineers find it easy to switch in and out from a mixing analysis state. Even after many years of mixing, they still find it possible to listen to a musical piece without thinking how long the reverb is, where the trumpet is panned to, or question the sound of the kick. Others simply cannot help it. This ability is very advantageous. Being able to listen to both the Macro (overview) and Micro (detail) elements will benefit the overall mix in the end.

Although it is far less enjoyable to analyze the technical aspects of a movie while watching it, it can make film students much better filmmakers. Sit, watch and learn how the masters did it—simple. The same approach works for mixing as well.

Every single mix out there, whether good or bad, is a lesson in mixing. Learning what is good or bad in mixing is just a matter of pressing play and actively listening to what has been done. Although mixing analysis cannot always reveal how every element was achieved, it can reveal much of what was done.

There are endless things to look for while analyzing other's mixes, and these can cover every aspect of the mix. Here are just a few questions you might ask yourself while listening:

- How loud instruments are in relation to one another?
- How instruments are panned?
- How do the different instruments appear laid-out on the frequency spectrum?
- How far are instruments with relation to one another?
- How much compression is applied on various instruments?
- Is there any automation that can be detected?
- How long are the reverbs?
- How defined are the instruments?
- How do different mix aspects change as the song advances?

A quick demonstration would be appropriate here. The following points provide a partial mixing analysis for the first 30 seconds of *Smells Like Teen Spirit*, by Nirvana, the album version:

- The tail of the reverb on the crunchy guitar is audible straight after the first chord (0:01).
- There is extraneous guitar noise coming from the right channel just before the drums are introduced (0:05).
- The crunchy guitar dives in level when the drums are introduced (0:07).
- Along with the power guitars (0:09–0:25), the kick on downbeats is louder than all other hits. It appears to be the actual performance, but it can also be achieved artificially during mixdown.
- When listening in mono, the power guitars lose some highs (0:09–0:25).
- The snare reverb changes twice (a particular reverb before 0:09, then no audible reverb until 0:25, then another reverb).
- During the verse, all the kicks have the same timbre suggesting drum triggers.
- There is kick drum reverb during the verse.
- It is possible to hear a left/right delay on the hi-hats—especially during open/close hits. This can be the outcome of a spaced microphone technique but could also be applied artificially during mixdown.
- The drums are panned audience view.

An excerpt set can be a true asset when it comes to mixing analysis as the quick changes between one mix to another make many aspects more noticeable. Not every aspect of the mix is blunt—some are felt rather than clearly heard. To be sure, the more time and practice we put into mixing analysis, the more we discover.

AN EXCERPT SET

For anyone learning how to mix, I suggest producing something called the excerpt set. It takes around half an hour to prepare but provides a lifetime mixing lesson. The excerpt set is very similar to a DJ set, only that each track plays for around 20 seconds, and we do not have to beat-match.

Simply pull around 20 albums from your CD library, pick only one track from each album and import all the different tracks into an audio sequencer or DAW. Then trim an excerpt of 20 seconds from each track and arrange the excerpts consecutively. It is important to balance the perceived level of all the excerpts, and it is always nice to have cross-fades between them. You are very likely to learn that mixes you thought were good are not as good when played before or after another mix. While listening, try to note mixes you think overpower others. Such an observation can promote a firmer endorsement to what a good mix is and later be used as a reference.

In addition to what we can hear from the plain mix, it is also possible to use different tools in order to reveal extra information. Muting one channel of the mix can disclose additional stereo information (e.g., a mono reverb panned to one extreme). Using a pass filter can help in understanding how things have been equalized. To unveil various stereo effects, one can listen in mono while phase-reversing one channel (this results in a mono version of the difference between the left and right, which tends to make reverbs and room ambience very obvious).

REFERENCE TRACKS

Mixing analysis is great, but it is impractical to "learn" hundreds of mixes thoroughly or carry them around just in case we want to refer to them. It is better to focus on a few selected mixes, learn them inside out, analyze them to the smallest detail of every aspect and have them readily accessible when needed.

Some mixing engineers carry a CD compilation with a few reference tracks (mostly their own past mixes, but can be standard "accepted" classic mixes) and upon occasion refer to them. The novice might refer to his reference tracks on a more frequent basis. Portable "MP3" players such as the iPod® are also used, ideally with the music stored in a lossless format. When mixing at home, some have a specific folder on the hard drive with selected mixes. Take along previous mixes of your material that you like or dislike for reference.

In addition to reference tracks, an excerpt set (see box above for details) can be great since it enables a quick comparison between many mixes. It is also possible to include a few raw tracks which can later be used to evaluate different tools.

Our choice of reference tracks might not be suitable for every mix. If we are working on a mix that includes strings, and none of our reference tracks involves strings, it would be wise to look for a good mix that does. Likewise, if our reference tracks are all heavy metal and we happen to work on a chill-out production, it would be wise to refer to a few chill-out mixes.

Usage of Reference Tracks

Reference tracks can be employed for different purposes:

- *As a source for imitation*: Painting students often go to a museum to copy a familiar painting. While doing so they learn the finest techniques of famous painters. Imitating others' techniques is all part of painting tuition—nothing invalid. Likewise, there is nothing invalid in imitating others' mixing techniques—if you like the sound of the kick in a specific mix, why not ask for it to be imitated in your mix? Why should you not craft your ambience just like the ambience on a specific track you like? When we are short of a mixing vision we can replace it with the sonic picture of an existing mix, try to imitate it or some aspects of it. Trying to imitate the sound of a known mix holds a great threat as well. First, productions are so diverse—whether in the emotional message, style, arrangement, quality and nature of the raw material, and so forth—that what sounds good in another mix might not sound good in yours. Second, setting a specific sound as an objective can mean that nothing better will be achieved. Finally, and most importantly, *imitation is innovation's greatest enemy*—there is hardly anything creative involved in imitation. In fact, it might restrain the development of creative mixing skills.
- *As a source for inspiration*: While imitating a mix requires a constant comparison between the reference track and our own mix, reference tracks can be played before mixing onset just to inspire us as to the direction of the mix and the qualities it should incorporate. For the novice such a practice can kick-start some mixing vision and set certain sonic objectives.
- *As an exit route from a creative dead end*: Sometimes we reach a point where we are clearly unhappy with our mix, but cannot tell what is wrong with it. We might be simply out of ideas or lacking any vision. Learning the difference between our mix and a specific reference mix can trigger new ideas or suggest problems in our mix.
- *As a reference to a finished mix*: When we finish mixing, we can compare our mix to a specific reference track. By simply listening to how other professionals did it, we can come up with ideas for improvements. The frequency response or relative levels of the two mixes are just two possible comparison aspects.
- *To calibrate your ears to different listening environments*: Working anywhere but in our native listening environment reduces our ability to evaluate what we hear. Just before we start to critically listen in unaccustomed environment, whether mixing or just evaluating our own mixes, playing a mix we know well can help in calibrating our ears to the unfamiliar speakers, the room itself or even a different position within the room.
- *To evaluate speaker models before purchase*: Studio monitor retailers usually play customers a known track, one that many people like, that has an impressive mix at loud levels. Chances are that the monitors will impress the

listener that way. However, doing so while listening to a mix we are actually fluent with can lead to better judgment, and in turn a better purchase.

Note: It is worth remembering that most reference tracks will have been mastered. Therefore, reference tracks are very likely to present tighter dynamics, usually in the form of more allied relative levels and heavier compression. In some albums, frequency treatment takes place in order to match the overall sound to that of the worst track. These points are worth bearing when comparing a reference track to a mix in progress—a mastered reference track is an altered version of a mix, mostly for the better. Therefore, the mix that comes out might not be at the same exacting standards of the mastered reference material. As such, be mindful that there is still one more important stage to go—mastering, which is covered by Bob Katz (see Chapters 6A–D).

How to Choose a Reference Track

Choosing some of our own past tracks for a reference track is very handy, simply because having worked on these tracks we know their finest details and their faults. Also, there is an advantage in referencing to an unmastered mix. Often, reference tracks are selected from commercial releases. Here are a few of the qualities that reference tracks should possess:

- *A good mix*: While being a matter of opinion, your opinion of what is a good mix counts first. It is important to choose a mix you like, not a production you like—with the greatest of respect to Elvis Presley, the sonic quality of his original albums are nowhere near today's standards.
- *A contemporary mix*: Mixing has always evolved. A good mix from the 1980s is likely to have more profound reverbs than the mix of a similar production from the 1990s. Part of the game is keeping up with the changing trends.
- *Genre related*: Clearly, it makes little sense to choose a reference track of a genre that is fundamentally different from the genres you will be working on.
- *Not a characteristic mix*: The mixing style of some bands, The Strokes for example, is tightly related to the style of music they play. A mix that has a highly distinct character will only serve those distinct productions and bands.
- *Not too busy*: It is usually easier to discern mixing aspects in sparse productions.
- *Not too simple*: The more there is to learn from a mix the better. An arrangement made of a singer and her acoustic guitar might sound great, but it will not help you assess a drum mix.
- *A dynamic production*: Choosing a dynamic production, which has a dynamic arrangement and a dynamic mix, can be like having three songs in one track. There is more to learn from such a production, including interest.

Below is one of my reference tracks. It has a rich, dynamic arrangement, which includes female and male vocals, acoustic and electronic guitars, keyboards, strings and varying drum sounds between sections. The mix presents a well-balanced frequency spectrum, high definition, beautiful ambience, musical dynamics and rich stereo panorama—all are retained through quiet and loud passages and changing mix densities.

"Witness" by The Delgadors from the Album "The Great Eastern," Chemikal Underground Records 2000.

FURTHER READING

This is an edited chapter from Izhaki, R. (2008). *Mixing Audio*. Focal Press, Oxford.

Chapter 5

B. The Process of Mixing

Roey Izhaki

In This Chapter

In this chapter, Roey Izhaki describes how the mix is started and how it progresses. An overview of the production process and how mixing is intrinsically linked is explored.

MIXING AND THE PRODUCTION PROCESS

Recorded Music

There are great differences between the production process of recorded music and sequenced music and how these affect the mixing process. Figure 5.3 shows the common production chain of recorded music as outlined in this book. Producers might have their input on each stage of the process, but they are mostly concerned with the arrangement (pre-production) and recording. Each stage has an impact on the stages succeeding it, yet different people can oversee each part.

Mixing is largely dependent on both the arrangement and the recordings. For example, an arrangement might involve only one percussion instrument such as a shaker. If the shaker is panned center in a busy mix, it is most likely to be masked by other instruments. But panning it to one side can create an imbalanced stereo image. It might be easier for the mixing engineer to have a second percussion instrument, say a tambourine, so the two can be panned left and right, respectively. A wrong microphone placement during the recording stage can result in the lack of body for the acoustic guitar. Recreating this missing body during mixdown is a challenge. Some recording decisions are, to be sure, mixing decisions. For example, the choice of stereo-miking technique for drum overheads determines the localization and depth of the various drums in the final mix. Altering these aspects during mixdown takes effort.

Mixing engineers, when being a separate entity in the production chain, are commonly facing arrangement or recording issues like the ones above. There is such a strong link between the arrangement, recordings and the mix, that it actually seems unreasonable for a producer or a recording engineer to have no mixing experience. A good producer foresees the mix. There is an enormous advantage in having a single person helping with the arrangement, observing the recording process and mixing the production. This assures that the mix is in the mind throughout the production process.

There is some contradiction between the nature of the recording and mixing stages. The recording stage is mostly concerned with capturing of each individual instrument as good as possible. During the mixing stage, different instruments have to be combined and their individual sound might not work perfectly well in the context of a mix. For example, the kick and bass might sound unbelievable when each is played in isolation but when combined they might mask one another. Filtering the bass might

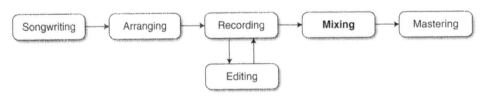

FIGURE 5.3
Common production process for recorded music.

make it thinner but will work better in mix context. Much of mixing involves altering recordings to fit into the mix—no matter how well instruments were recorded.

Sequenced Music

The production process of sequenced music (Figure 5.4) is very different in nature to that of recorded music. In a way, sequenced music production is regarded as a mishmash between song-writing, arranging and mixing. This affects mixing in two principal ways. First, today's digital audio work-stations, on which most sequenced music is produced, make it easy to mix as-you-go. The mix is an integral part of the project file, unlike a console mix, which is stored separately from the multitrack. Second, producers commonly select samples or new sounds while the mix is playing along; unconsciously, they choose sounds based on how well they fit into the existing mix. A specific bass preset might be dismissed if it lacks definition in the mix, and a lead synth might be chosen based on the reverb that comes along with it. Some harmonies and melodies might be transposed so they blend better into the mix. The overall outcome of this is that sequenced music arrives to the mixing stage partly mixed.

FIGURE 5.4
Common production process for sequenced music.

As natural and positive this practice might seem, it promotes a few mixing problems that are typical to sequenced music. First, synthesizer, manufacturer and sample-library publishers often add reverb (or delay) to presets in order to make them sound bigger. These reverbs are ironclad into the multitrack submission and have restricted depth, stereo image and frequency spectrum that might not integrate well with the mix. Generally speaking, dry synthesized sounds and mono samples give more possibilities during mixdown. Second, producers sometimes get attached to a specific mixing treatment they have applied, like the limiting of a snare drum, and leave these treatments intact. Very often the processing is done using inferior plug-ins, in a relatively short time, and with very little attention to how the processing affects the overall mix. Flat dynamics due to over-compression or ear-piercing highs are just two issues that might have to be rectified during the separate mixing stage.

Recording

They say that all you need to get the killer drum sounds is a good drum kit in a good room, fresh skins, a good drummer, good microphones, good preamps, some good EQs, nice gates, nicer compressors and a couple of good reverbs. Take one of these out and you will probably find it harder to achieve that killer sound; take three and you might never achieve it. All these aspects need to be considered before recording to ensure quality is sustained throughout.

The quality of the recorded material has an enormous influence on the mixing stage. A famous saying is "garbage in : garbage out." Flawed recordings have to be rectified during mixing and often there is much we can rectify. Good recordings leave the final mix quality to the talent of the mixing engineer and offer greater creative opportunities.

Nevertheless, experienced mixing engineers can testify how drastically the process of mixing can improve poor recordings, and how even low-budget recordings can turn into an impressive mix. Much of this is thanks to the time, talent and passion of the mixing engineer.

Arrangement

The arrangement (or instrumentation) largely determines which instruments play, when and how. Mixing-wise, the most relevant factor of the arrangement is its density. A sparse arrangement (Figure 5.5a) will call for a mix that fills various gaps in the frequency, stereo and time domains. An example for this would be an arrangement based solely on an acoustic guitar and one vocal track. The mixing engineer's role in such a case is to create something out of very little. On the other extreme is a busy arrangement (Figure 5.5b), where the challenge is to create a space in the mix for each instrument. It is harder to protrude a specific instrument or emphasize fine details in a busy mix. Technically speaking, masking is the cause.

As such there is an argument to suggest that mixing starts with the composer and arranger of the piece and not with the mix engineer. Consider a classical orchestration. Its mix will be the responsibility of the composer armed with the ability to orchestrate and convey his vision using the score and of the conductor who ultimately will "produce" how the performance will be delivered.

Therefore, it is important to have a vision of what you wish for early on in the genesis of the music. If you are arranging the piece, then consider the elements that will work in the mix later on at an early stage. Clever arrangements with innovative production can often lead to exciting and energetic performances and ultimately great mixes.

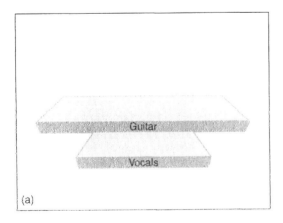

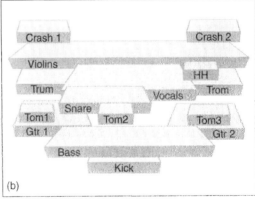

FIGURE 5.5
Busy vs. sparse arrangement.

©2009 Roey Izhaki

SOME AUDIO EXAMPLES

Polly
Nirvana. *Nevermind* [CD]. Geffen Records, 1991.
Mixed by Andy Wallace.

Exit Music (For a Film)
Radiohead. *OK Computer* [CD]. Parlophone, 1997.
Mixed by Nigel Godrich.

Hallelujah
Jeff Buckley. *Grace* [CD]. Columbia Records. 1994.
Mixed by Andy Wallace.

Both Andy Wallace and Nigel Godrich faced a sparse arrangement, made of a guitar and vocal only, in sections of Polly and Exit Music. Each tackled it in a different way—Wallace chose a plain intimate mix, with fairly dry vocal and a subtle stereo enhancement for the guitar. Godrich chose to use very dominant reverbs on both the guitar and vocal. It is interesting to note that Wallace has chosen the latter reverberant approach on his inspiring mix for Hallelujah by Jeff Buckley—a nearly 7-minute song with an electric guitar and a single vocal track.

It is not uncommon for the final multitrack to include extra instrumentation along with takes that were recorded as tryouts in order to give some choices during mixdown. It is possible, for example, to get with one song, eight power-guitar overdubs. This is done with the belief that layering eight takes of the same performance will result in enormous sound. Enormousness aside, properly mixing only two tracks out of the eight can sometimes sound much better. There are always opposite situations where the arrangement is so minimalist that it is very hard to cast a rich, dynamic mix. In such cases, nothing should stop the mixing engineer from adding instruments to the mix—as long as time, talent and ability allow this, and the client approves the additions.

It is worth remembering that the core process of mixing involves both alteration and addition of sounds—a reverb, for example, is an additional sound that occupies space in the frequency, stereo and time domains. It would therefore be perfectly valid to say that a mix can add to the arrangement. Some producers take this well into account by "leaving a place for the mix"—we might associate the famous vocal echo on Pink Floyd's *Us and Them* as such affair. To a greater extent, a specific production philosophy adheres to rather simple arrangements, which are then turned potent by a powerful mix. Mixing in such cases is a dominant means of production.

Editing

On many projects that are not purely sequenced, editing is the final stage before mixing. Editing is subdivided into two types: selective and corrective. Selective editing is primarily concerned with choosing the right takes and the practice of comping (combining multiple takes into a composite master take). Corrective editing is done to repair a bad performance. Anyone who has ever engineered or produced in a studio knows that session and professional musicians are a true asset. But as technology is moving forward, enabling more sophisticated performance corrections, poor performance is becoming more accepted. Why should we spend money on vocal tuition and studio time, when a plug-in can make the singer in tune? (More than a few audio engineers believe that the general public perception of pitch has sharpened in recent years due to the excessive use of pitch-correction.) Drum correction has also become a common practice. On big projects a dedicated editor (perhaps the Pro Tools operator from the recording sessions) might be working with the producer to do this job. Unfortunately though, very often it is the mixing engineer who completes these tasks.

Corrective editing can be done to a mechanical extent. Most drums can be quantized to a metronomic precision and vocals can be made perfectly in tune. Although some pop albums feature such extreme edits, many advocate a more humanized approach, which calls for little more than acceptable performance (perhaps ironically, sequenced music is often humanized to give it feel and swing). Some argue that over-correcting is unlawful to genuine musical values. It is also worth remembering that corrective editing always involves some quality penalty. In addition, audio engineers are much more sensitive to subtle details than most listeners. To give one example, the chorus vocals on Beyoncé Knowles' *Crazy in Love* are notoriously late and offbeat, many listeners don't notice it.

THE MIX AS A COMPOSITE

Do individual elements constitute the mix, or does the mix consist of individual elements? Those who believe that individual elements constitute the mix might give more attention to how the individual elements sound (Micro), but those who think that the mix consists of individual elements would care about how the sound of individual elements contribute to the overall mix (Macro). It is worth remembering that the mix, as a whole, is the final product. This is not to say that the sound of individual elements is not important, but the overall mix has priority.

A few examples would be appropriate here. It is extremely common to apply a high-pass filter on a vocal in order to remove muddiness and increase its definition. This type of treatment, which is done in various degrees, can sometimes make the vocals sound utterly unnatural, especially when soloed. However, this unnatural sound often works extremely well in mix context. Another example: Vocals can be compressed while soloed, but the compression can only be perfected when the rest of the mix is playing as well—level variations might become more noticeable with the mix as a reference. Overhead compression should also be evaluated against the general dynamics and intensity of the mix.

It even goes into the realm of psychoacoustics—our brain can separate one sound from a group of sounds. So, for example, while equalizing a kick we can isolate it from the rest of the mix in our heads. However, we can just as well listen to the whole mix while equalizing a kick and by that improving the likelihood of the kick sounding better in mix context. This might seem a bit abstract and unnatural, while we manipulate something we seek to clearly hear the effect. The temptation to focus on the manipulated element always exists, but there's a benefit in listening to how the manipulation affects the mix as a whole.

WHERE DOES THE MIX ENGINEER START?

Preparations

Project files submitted to the mixing process are usually accompanied by documentation such as session notes, track sheets, edit notes and others. These should all be inspected before mixing commences. Clients often have their vision, ideas, guidelines or requirements regarding the mix, which should often be discussed at this stage. In addition, there are various technical tasks that might need to be accomplished; these are discussed shortly.

Auditioning and the Rough Mix

Unless the engineer was involved in earlier production stages and thus fluent with the raw tracks, he or she must listen to what is about to be mixed. Auditioning the raw tracks lets them learn the musical piece, capturing its mood and emotional context, identifying important elements, moments or problems they have to rectify. The engineer must learn the ingredients before they start cooking.

Often a rough mix (or a monitor mix) is provided with the raw tracks. This can teach much about the song and inspire a particular mixing direction. Even when a rough mix is submitted, creating one can also be extremely handy—in doing so the engineer investigates the arrangement, structure, quality of recording and, maybe most importantly, how the musical qualities can be conveyed. This personal rough mix is a noncommittal chance to learn much of what they are dealing with, what has to be dealt with and how. It also helps us all to create a mixing plan or vision.

One issue with rough mixes is that both the artist and production team can, sometimes unwontedly, get used to them and embrace them as a paragon for the final mix. This unconscious adoption is only natural since a rough mix often provides the first chance to see how many elements turn into something that starts to sound like the real thing. An exciting moment indeed.

However, this adoption can be dangerous as rough mixes, by nature, are done with little attention to small details, and sometimes even involve random tryouts that make little technical or artistic sense. Yet, once used to them, we all find it hard to live without them or their facets—a point worth remembering.

The Plan

Just bringing faders up and doing whatever seems right surely won't be effective. It's like playing a football match without tactics. Every mix is different and different pieces of music require different approaches. Once engineers are familiar with the raw materials a specific plan, a course of action, a vision can be written either in mind or on paper. Such a plan can help even when the mixing engineer recorded or produced the musical pieces—it resets the mind from any sonic prejudgments and sets a fresh start to the mixing process. Below is just one example of what might be the beginning of a rough plan before mixing starts. As can be seen, this plan includes various ideas, from panning positions to the actual equipment to be used. There are virtually no limits to what such plan might entail and in what format:

> I am going to start by mixing the drum-beat at the climax section. In this production the kick should be loud and in-your-face; I'll be compressing it using the Distressor, accent its attack, then add some sub-bass. The pads will be panned roughly halfway to the extremes and should sit behind everything else. I should also try and distort the lead to tuck on some aggression. I would like it to sound roughly like Newman by Vitalic.

It might be hard to get the feel of the mix at the very early stages. Understandably, it is impossible to write a plan that includes each and every step that should be performed. Moreover, such a detailed plan can limit creativity and chance, which are important aspects of mixing. Therefore, instead of one big plan it can be easier to work using *small plans*. Whenever a small plan is finished, a new evaluation of the mix takes place and a new plan is established. Here is a real-life example of a partial task list from a late mixing stage:

- Kick sounds flat
- Snare too far during the chorus—replace with triggers (chorus only)
- Stereo imbalance for the violins—amend panning
- Solo section: violin reverb is not impressive enough
- Haas guitar still not defined
- Automate snare reverbs.

Not every mixing engineer approaches mixing with a detailed plan, some do it unknowingly in their heads. Yet, there is always some methodology or some procedure being followed.

Technical vs. Creative

The mixing process involves both technical and creative tasks. Technical tasks are usually ones that do not affect the sounds or ones that relate to technical problems with the raw material. They usually require little sonic expertise. Here are a few examples:

- *Neutralizing (resetting) the desk*: Analog desks should be neutralized at the end of each session but nature has it that this is not always the case. Line gains and aux sends are the usual suspects that can later cause troubles. Line gains can lead to unbalanced stereo image or unwanted distortion. Aux sends can result in unwanted signals being sent to effect units.
- *Housekeeping*: Projects might require some additional care, sometimes simply for our own convenience. For example, files renaming, removing unused files, consolidating edits and so on.

- *Track layout*: Organizing the appearance order of the tracks, so they are convenient to work with. For example, making all the background vocals, or drum tracks consecutive in appearance, or having the suboscillator next to the kick. Sometimes the tracks are organized in the order by which they will be mixed, and sometimes the most important tracks are placed in the center of a large console. Different mixing engineers have different layout preferences to which they usually adhere. This enables faster navigation around the mixer and increased accessibility.
- *Phase check*: Recorded material can suffer from various phase issues. Phase problems can have subtle to profound effects on sounds, and it is therefore important to deal with them at the beginning of the mixing process.
- *Control/audio grouping*: Setting any logical group of instruments as control or audio groups. Common groups are drums, guitars, vocals etc.
- *Editing*: Any editing or performance correction that might be required.
- *Cleaning up*: Many recordings require cleaning up unwanted sounds like the buzz from guitar amplifiers. Cleaning up can also include the removal of extraneous sounds like count-ins, musician talks, pre-singing coughs etc. These unwanted sounds are often filtered either by gating, a strip-silence process, or by region trimming.
- *Restoration*: Unfortunately raw tracks can incorporate noise, hiss, buzz or clicks. These are more common in budget recordings but can also appear due to degradation issues. It is important to note that clicks might be visible but not audible (inaudible click can become audible if being processed by devices like enhancers or if being played through different D/A converters). Some restoration treatment, de-noising for example, might be applied to solve these problems.

Creative tasks are essentially the ones by which we craft the mix. These might include:

- Using a gate to shape the timbre of a floor tom.
- Tweaking a reverb preset to sweeten a saxophone.
- Equalizing vocals to give them more presence.

While mixing, engineers have a flow of thoughts and actions. The creative process, which consists of many creative tasks, is an attentive one, and it usually requires a high degree of concentration. Any technical task can distract or break the creative flow. Whilst equalizing a double bass the engineer may find that it is offbeat on the third chorus and might be tempted to fix the performance straight away (after all, a bad performance can be highly disturbing). By the time the editing is done, the engineer might have switched-off from the equalizing process or the creative process altogether. It can take some time to get back to creative mood. It is therefore beneficial to go through all the technical tasks first, which clears the path to a distraction-free creative process.

Which Instruments?

Different engineers each have a different order in which they mix the component tracks. Some are not committed to one order or another—each production might be mixed in the order they think is most suitable. Here is a summary of common approaches, their possible advantages and disadvantages:

- *The serial approach*: Starting from a very few tracks, we listen to them in isolation and mix them first, then gradually more and more tracks are added and mixed. This divide-and-conquer approach enables the engineer to focus well on individual elements (or stems). The danger is that as more tracks are introduced there is less space in the mix.
  - *Rhythm, harmony, melody*: Starting by mixing the rhythm tracks in isolation (drums, beat and bass), then other harmonic instruments (rhythm guitars, pads and keyboards) and finally the melodic tracks (vocals and solo instruments). This method often follows what might have been the overdubbing order. It can also feel a bit odd to work on drums and vocals without any harmonic backing. But arguably, from mixing point of view, it makes little sense mixing an organ before the lead vocals.
  - *By the order of importance*: Tracks are brought up and mixed by the order of importance. So, for instance, a hip-hop mix might start with the beat followed by lead vocals, additional vocals and then all the other tracks. The advantage here is that important tracks are mixed at early stages when there is still space in the mix and so they can be made bigger. The least important tracks are mixed last into a crowded mix, but there is less of a penalty in making them smaller.
- *Parallel approach*: This approach involves bringing all the faders up, setting a rough level balance, rough panning and then mixing individual instruments in whatever order one desires. The advantage with such an approach is that nothing is being mixed in isolation. It can work well with small arrangements but can be problematic if many tracks are involved—it can be very hard to focus on individual elements or even make sense of the overall mix at its initial stages. By way of analogy, it can be like playing football with eight balls on the pitch.

There are endless variations to each of these approaches. Some, for example, start by mixing the drums (the rhythmical spine), then progress to the vocals (most important), then craft the rest of the mix around these two important elements. Another approach, which is more likely to be taken when an electronic mix makes strong usage of the depth field, involves mixing the front instruments first, then adding the instruments panned to the back of the mix.

There are also different approaches to *drum mixing*. Here are a few things to consider:

- *Overheads*: The overheads are a form of reference to all the other drums. For example, the panning position of the snare might be dictated by its position on the overheads. Changes made to the overheads might affect other drums; so there is some advantage in mixing them first. Nonetheless, many engineers prefer to start from the kick followed by the snare and only then they might mix the overheads. Sometimes the overheads are even the last drum track to be mixed.
- *Kick*: Being the most predominant rhythm element in most productions, the kick is often mixed before any other individual drum and sometimes even

before the overheads. Following the kick, the bass might be mixed and only then other drums.

- *Snare*: Being the second most important rhythm element in most productions, the snare is often mixed after the kick.
- *Toms*: Toms often only play occasionally, which makes them somewhat the least important contributors to the overall sound of the drums. Yet, their individual presence in the mix can be highly important.
- *Cymbals*: The hi-hats, ride, crashes or any other cymbals might have a sufficient presence in the overheads. Often in such cases, the cymbals are used to support the overheads or only mixed at specific sections of the song for interest sake. Sometimes these tracks are not mixed at all.

Which Section?

With rare exceptions, the process of mixing involves working *separately on the various sections*. Each section is likely to involve different mixing challenges and a slightly different arrangement (choruses are commonly denser than verses). And so, mixing engineers usually loop one section, mix it, then move on to the next section and mix it based on the existing mix. The question is: Which section should be first? There are two approaches here:

- *Chronologically*: Starting from the first section (intro) and slowly advancing to succeeding sections (verse, chorus). It seems very logical to work this way as this is the order by which music is played and recorded. However, while we might mix the verse to the best extent—creating a rich and balanced mix—there will be very little place in the mix for new instruments introduced during the important chorus.
- *By the order of importance*: The most important section of the song is mixed first, followed by the less important sections. For a recorded production, this section is usually the chorus; for some electronic productions it will be the climax. Very often, these most important sections are also the busiest ones; therefore, mixing them first can be beneficial.

Which Treatment Should Be Applied First?

The standard guideline for treatment order is shown in Figure 5.6. With the exception of faders, which need to be up for sound to be heard (unless soloed), there is nothing wrong in not complying with this order. There is some logic in the order above. If the engineer skips panning and mixes in mono for a while, both the width and depth dimensions are lost. As masking is most profound in mono, a few engineers choose to resolve masking by equalization while listening in mono. This might be done before panning but more often done using mono summing. Since processing replaces the original sound, it comes before any effects that add to the sound (modulation, delay or reverb). The assumption is that it is preferred to have the processed signal sent to the effects, rather than what might be a problematic unprocessed signal. On nearly the same basis, it is usually desired to have a modulated sound delayed, rather than having the delays modulated. Finally, as

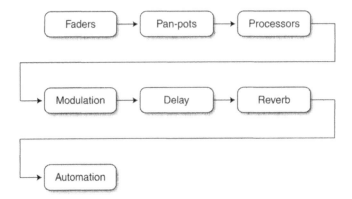

reverb is generally a natural effect, it is normal to have it untreated; treated reverbs are considered as creative effect. In many cases, automation is the last major stage in a mix.

There is also the dry–wet approach, in which all the tracks are first mixed using dry treatment (faders, pan pots and processors) and only then wet treatment is applied (modulation, delay and reverb). This way, the existing sounds are all dealt with before adding new sounds to the mix. It also leaves any depth or time manipulations (reverbs and delays) for a later stage, which can simplify the mixing process for some. However, some claim that it is very hard to get the real feel and the direction of the mix without depth or ambience.

Finally, it should be said that the very last stage of the mix, before it is printed, usually involves refinements of panning and levels.

The Iterative Approach

Back in the days of two-, four- and eight-track tape recorders, mixing was an integral part of the recording process. For example, engineers used to record drums onto six tracks, then mix and bounce them to two tracks and use the previous six tracks for additional overdubs. The only thing that limited the amount of tracks to be recorded was the accumulative noise added in each bounce. Back then, engineers had to commit their mix time and again throughout the recording process. Once bouncing took place and new material wrote over the previous tracks, there was no way to revert to the original drum tracks. Such a process required enormous forward planning from the production team—they had to mix considering a relation to something that was not even recorded yet; diligence, imagination and experience were the key.

Today's technology offers hundreds of tracks. Even when submix bouncing is needed (due to channel shortage on a desk or processing shortage on a DAW), the original tracks can be reloaded at later times and a new submix can be made. This practically means that everything can be altered at any stage of the mixing process and nothing has to be committed before the final mix is printed.

Flexibility to amend mixing decisions at any point in the mixing process is a great asset since mixing is a highly correlated process. First, the existing mix should normally be retouched to accommodate newly introduced tracks. For example, no matter how good the drums sound when mixed in isolation, the introduction of distorted guitars into the mix might require additional drum treatment (the kick might lose its attack, the cymbals definition and so forth). Second, any treatment in one area might require subsequent treatment somewhere else. For instance, when brightening the vocal by boosting the highs, high frequencies may linger on the reverb tail in an unwanted way; so the damping control on the reverb might need to be adjusted. The equalization might also make the vocal seem louder in the mix and so the fader might need to be adjusted. If the vocal signal is equalized before it is sent to the compressor, it is likely the compression's settings will need to be altered accordingly.

As we've previously mentioned, mixing is such a correlated process and as such it can benefit from an iterative coarse-to-fine approach (Figure 5.7). Engineers start with coarse treatment on which they spend less time, then as the mix progress they refine previous mixing decisions. Most of their attention is given to the late mixing

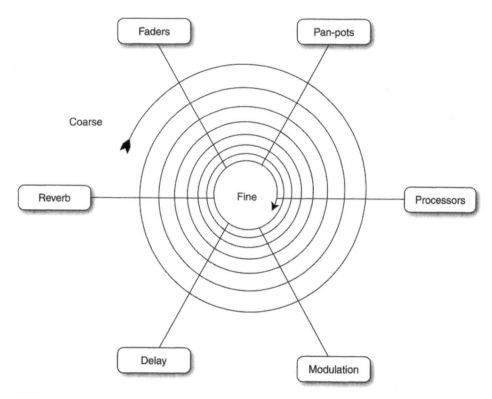

FIGURE 5.7
The iterative coarse-to-fine mixing approach.

stages where the subtlest mixing decisions are made. There is little justification in trying to get everything perfect before these late stages—what is perfect at one point might not be that perfect later.

DEADLOCKS

The Evaluation Block

This is probably the most common and frustrating deadlock for the novice. It involves listening to the mix, sensing that something is wrong, but not being able to tell what. Reference tracks are a true asset in these situations—they give an opportunity to compare aspects of the mix to an established work. This comparison can reveal the wrongs in the mix or at least give a direction as for how things can be made better.

The Circular Deadlock

As engineers mix they define and execute various tasks—from small tasks like equalizing a shaker to big tasks like solving masking or crafting the ambience. The problem is that in the process they naturally tend to remember their most recent actions. While at any given moment an engineer might not pay any attention to a specific compression applied 2 days before, he or she might be tempted to reconsider an equalization applied an hour before. Thus, it is possible to enter a frustrating deadlock in which recent actions are repeatedly evaluated. Instead of pushing the mix forward, the mix goes round in circles. An easy way out of this situation could be a short break. The production team, perhaps with the artist also present, might also want to listen to the mix as a whole and re-establish a plan based on what really needs to be done.

The Raw Tracks Factor

Life has it that sometimes the quality of the raw tracks are poor to such an extent that they are impossible to work with. For instance, distortion guitars that exhibit strong comb filtering will often take blood to fix. A known saying is: What cannot be fixed, can be hidden, trashed or broken even more. For example, if an electric guitar was recorded with a horrible pedal monophonic reverb that just does not fit into the mix, maybe over-equalizing it and ducking it in relation to a delayed version of the drums will yield such an unusual effect that nobody would notice the mono reverb anymore. Clearly, there is a limit to how many instruments can receive such prestige treatment, so sometimes re-recording is inevitable. In other cases, it is the arrangement to blame, like when the recorded tracks involve a limited frequency content or the absence of a rhythmical backbone. Again, there is nothing to stop the mixing engineer from adding new sounds or even re-recording instruments; nothing, apart from time availability and ability, plus permission from the artist or producer.

MILESTONES

The mixing process can have many milestones. On the Macro level we can define a few key milestones. The first milestone involves bringing the mix into an adequate state. Once this milestone is reached, the mix is expected to be free of any issues—whether those existed on the raw tracks or those that were introduced during the actual process of mixing. Such problems can span from basic issues like relative levels (e.g., solo guitar too quiet) to more advanced concepts like untidy ambience (Figure 5.8).

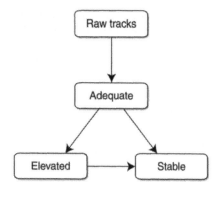

FIGURE 5.8
Possible milestones in the mixing process.

Nevertheless, a problem-free mix is not necessarily a good one, the next step would be to elevate the mix to distinction. The definition of a distinctive mix is abstract and varies in nature between one mix to another, but the general objective is to make the mix notable—something to remember—whether by means of interest, power, feel or any other sonic property.

Finally, a step that might be reserved to the inexperienced engineer only is stabilizing the mix. This step is discussed in detail in the next section.

FINALIZING AND STABILIZING THE MIX

All the decisions made during mixing are based on the evaluation of what the engineer hears at the listening position, otherwise known as the sweet spot. But mixes can sound tragically different once one leaves this sweet spot. Here are prime examples:

- Different speakers or as they're called in the trade—"monitors," reproduce mixes differently—each speaker model (in combination with its amplifier) has its own frequency response, which also affects perceived loudness. While listening on a different set of speakers the vocals might not sound as loud and the hi-hat might sound harsh. Different monitor positions also affect the overall sound of the mix, mainly with relation to aspects like stereo image and depth.
- Mixes sound different in different rooms—each room has its own sonic stamp that affects the music played within it. Frequency response, which is affected by both room modes and comb filtering, is the most notable factor, but the room response (reverberation) also plays a part.
- Mixes sound different in different points in a room—both room modes and comb filtering alter our frequency perception as one moves around the room. A mix will sound different when stood next to the wall than when stood in the middle of the room.
- Mixes sound different at different playback levels too.

Many people are familiar with the situation where their mix sounds great in their home studio but translates badly when played in their car. It would be impossible to have a mix sounding the same in different listening environments for the reasons explained above. This is worth repeating: mixes *do* sound very different in different places. What should be ensured is that the mixes are problem-free in other listening environments. This is much of the idea behind stabilizing the mix.

The reason that it is usually the novice or amateur that has to go through this practice is that the veteran mixing engineer knows his mixing environment so well that he can predict how well the mix will sound in other places. Also, in comparison to a professional studio, the sweet spot in a home studio is often quite sour. Acoustic problems in home studios can be very profound. Meanwhile, in a professional studio these are rectified by expensive construction and acoustic design carried out by experts.

The process of stabilizing the mix involves listening to the mix in different venues, different positions in these venues and at varying levels. Based on what we all learn from this practice we can finalize the mix by fine-tuning some of its aspects. Normally this involves subtle level and equalization adjustments. Here is how and where mixes should be listened to during the stabilizing phase:

- *Quiet levels*: The quieter the level, the less prominent the room becomes and any deficiencies it embraces. Listening quietly also alters the perceived loudness of the lows and highs, and it is always a good sign if the relative level balance in the mix hardly changes as the mix is played at different levels. Listening at loud levels is also an option but can be dubious as mixes most often sound better when played louder.
- *At the room opening*: Also known as "outside the door," many find this listening position highly useful. When listening at its opening, the room becomes the sound source, and instead of listening to highly directional sound coming from the (relatively) small speakers we listen to reflections that come from many points in the room. This reduces many acoustic problems caused by the room itself, for example, comb filtering. The monophonic nature of this position can also be an advantage.
- *Car stereo*: A car is a small, relatively dry listening environment that many people find appealing. Same like with the room-opening position, it provides a very different acoustic environment that can reveal problems.
- *Headphones*: With the growing popularity of "MP3" players comes the growing importance of checking mixes on headphones. The idea of headphone mix edits for digital downloads seems more and more logical nowadays. Headphones sacrifice some aspects of stereo image and depth for the absence of room modes, acoustic comb filtering and left/right phase issues. We can regard listening in headphones as listening to the two channels in isolation, and it is a known fact that headphones can reveal noises, clicks or other types of problems that would not be as noticeable when the mix is played through speakers. The main issue with headphones is that most of them have an extremely uneven frequency response, with strong highs and weak lows. Therefore, these are not regarded as a critical listening tool.
- *Specific points in the room*: As previously mentioned, many home studios can present poor acoustic behavior and very often there are issues at the sweet spot. Moving out of this spot can result in an extended bass response and many other levels or frequency revelations. Also, the level balance of the mix

changes as we move off axis and out of the monitors' close proximity range. Beyond a certain point, known as the critical distance (where the balance of the direct sound and the reverberant sound are equal), the room's combined reflections become louder than the direct sound. Most end-listeners hear mixes exactly under such conditions, so it is important to see what happens away from the sweet spot.

Evaluating the mix using any of these ways can be misleading if a caution is not exercised. When listening in a specific point in a room, it might seem that the bass disappears and subsequently we might be tempted to boost it in the mix. But there is a chance that it is that point in the room causing the bass deficiency, whereas the mix itself is fine. How then shall we trust what we hear in any of these places? The answer is that we should focus on specific places and learn the characteristics of each. Usually, rather than going to random points, people evaluate mixes exactly in the same point in the room every time. In addition, playing a familiar reference track to see how it sounds in each of these places is a great way to learn the sonic nature of each place.

Whilst listening to the mix in different places, it would be wise to write down what issues you notice need further attention. With those comments on board, take them to the studio and discuss with the production team. In some cases, comments cancel out one another. For example, if when using headphones it seems that the cymbals are too harsh, but when listening at the room opening they sound dull, perhaps nothing should be adjusted.

As some final advice, it should be said that listening in different places is not reserved for final mix stabilization—it can also be beneficial at other stages of the mixing process, especially when decisions seem hard to make.

FURTHER READING

This chapter is an edited extract from Izhaki, R. (2008). *Mixing Audio*. Focal Press, Oxford.

Gibson, D. (2005). *The Art of Mixing: A Visual Guide to Recording, Engineering and Production*, Aristopro.

Owinski, B. (2008). *The Mixing Engineer's Handbook*. Course Technology.

Stavrou, M. (2003). *Mixing with Your Mind*. Flux Research, Australia.

White, P. (2003). *Creative Recording, Part 1: Effects and Processors*. Sanctuary Publishing.

www.mixingaudio.com.

www.mixonline.com.

www.soundonsound.com.

Chapter 6

A. What Is Mastering?

Russ Hepworth-Sawyer

In This Chapter

Mastering is the skill of modifying the sound of a mix to give the best reproduction possible via its delivery format (CD, Vinyl, etc.) and the service to produce the final master(s) from which all the manufactured copies will be made. It is also the last opportunity for any artistic input by those involved in the production so the mastering engineer should be sympathetic to the sonic quality that the producer and artist wish to achieve. Processing may be required for a variety of reasons: Mixes may have been made in a poor acoustic environment or with sub standard monitoring; Optimum use may not have been made of the studio equipment and occasionally recordings have been poorly engineered. The fun part is the creative processing to increase the punch, impact and energy or sweeten the sound. That final touch that can make the recording more enjoyable to listen to.

Ray Staff (legendary mastering engineer)

INTRODUCTION

At the end of the mixing session, the studio should provide the mix on whatever format you require. Some mix engineers still prefer to use ½ inch analog tape for the analog factor but this is happening less and less. Digital audio tape (DAT) is still considered a safe format and is still used now and then. It is most likely though that the music will leave the studio on a CD-R, DVD-R or hard drive containing data files such as a .WAV file or similar. What now? Well this music has one more creative path before it is all over—Mastering.

MASTERING WAS . . .

Well *is* a process called pre-mastering to be exact. Historically speaking, the mastering stage, as we understand it today, was not to necessarily improve the sonic characteristics of the material. Its main job was to match the material for the format it was to be duplicated to. For example, if the music was to be pressed onto vinyl, as most records were for so many years, the mastering engineer would need to equalize the work for that medium, ensuring that the bass was managed and the level would also need to be managed, so that signals did not break through walls to the next groove when cutting for example.

These skills are still out there and many of the major mastering houses still retain their lathes, or disc cutters, for this exact purpose! However, mastering as a purpose has shifted over the years gradually to a process that not only ensures that it makes the medium play accurately but it also "improves" and "polishes" the sonic qualities of the music.

"In the early years disc cutting engineers often worked with a minimal amount of equalization or compression. They were essentially transfer engineers. Many cutting engineers believed that the mix should be cut as flat as possible and the producer and artist were sometimes not even allowed to attend the cut. This prompted a lot of effort from some recording engineers to ensure their work was right when it left the studio. The idea that it can be fixed in the cut simply did not exist. Many engineers who trained at major studios had to spend a period of time cutting records before they were allowed to progress to recording engineer status, this was invaluable experience for them. From around the end of the sixties mastering become far more creative. Independent cutting rooms started to provided a creative mastering environment. As each new format has arrived (cassette, 8 track cartridge, CD and compressed audio for downloads) we have had to evolve into the high quality mastering services you see today."

Ray Staff

MASTERING IS . . .

Mastering is now considered the last creative stop on the train line to the end of a production project. Whilst it is far from the end of the line, as this book I hope demonstrates, it is indeed the last station stop at which the audible content can be altered, improved, shaped and prepared for market.

Mastering is often considered a Jedi art of audio engineering because of its ability to improve and manipulate the sonic quality of material with relative ease. It is considered to be a secretive art form for good reason. Many recording and mix engineers have the same equipment (more or less) and the same ability to equalize, compress, limit and edit. This is not disputed, as these skills are more or less transferable. The art is in the skill of mastering, not the equipment.

Mastering Is . . . in Part, Listening

Where the mastering engineer differs to that of recording and mixing engineers is in the type of listening skills. As has been discussed by authors within this book, there are overarching listening types to note: Macro listening and Micro listening.

Micro listening is the ability to work with small elements of the mix, as a mixing engineer would, and scrutinize each element in fine detail, perhaps soloing the part in question. Taking into account each instrument, its musical content and sonic reproduction. Micro listening is essential for all engineers in all walks of life; however, some will do this more often than others and for different reasons. A recording engineer will focus on the instrumentation being recorded to ascertain that it has been played properly, it is in tune and the best performance has been captured. The mastering engineer, however, will use this ability (listening to an already completed stereo mix) to focus on elements within the mix that need scrutiny, such as a sibilant vocal, an errant noise or anything else for that matter.

Macro listening is the ability to objectively view the whole broadband audible content as one. To listen in this overarching way is an excellent skill that mix engineers and mastering engineers perhaps focus on that little bit more as the "whole" is their product. Listening in this way is a form of detachment that is important for the mastering engineer to learn so that an even hand can be applied to the music.

The "secretive" art mantle has been levied because of this detachment and the mastering engineer's ability to preside over the album and consider it as one "whole" entity. Many see the mastering process as quality control, ensuring that their mixes are in order and to gain comments from those whose opinion they might trust that it is indeed "a good mix."

WHAT CAN BE DONE?

A Look at the Processes

As we have established, mastering contains a number of processes that could be described as being quite similar to mixing. These processes may be common to the recording studio but perhaps are different for differing reasons in the mastering suite and can be of the highest quality possible. These processes include equalization, compression, limiting and any other special approaches that can be used.

A number of processes are applied to each track to lift its audible qualities to the best possible standards. A mastering engineer will have preferred tools and techniques that bring out the best in a piece of recorded material. Many consider this the "sweetening" of the music through the use of EQ and other tools such as dynamics processing that is introduced later in this book.

All this processing will not only improve each track but will also improve each one in relation to the rest of the album to ensure parity. Therefore, the quality of the sound is retained across the whole work as well as being ordered in such a way that it retains listener interest. The order is something that the artist, producer and label will usually decide. The mastering engineer will sequence or assemble the audio into a shape that makes it "flow" when played on a CD player. Bob Katz introduces the importance that the order of the music places on the flow of an album (Chapter 7). He elaborates to speak about the way in which material works together and if it doesn't, how it might be made to flow.

In this book, we do not delve too far into the art of mastering and processing as you will see. These chapters are intended as a guide to what goes on in the mastering suite and for an appreciation of the "art and science," that is, Mastering (Katz, 2007).

The Codes

PQ CODES

At the mastering suite, the engineer will also be very concerned about logging and details. Mastering is not only a creative role, but it also encompasses some considerable administration. It is the engineer's duty to provide the manufacturer with what is known as a PQ sheet or listing. These lists detail the P and Q flags in the CD data. The P flag simply informs the CD player that a new track is following; whilst the Q flag contains data about the track's length and duration and codes that give the track its identity....

INTERNATIONAL STANDARD RECORDING CODE

The International Standard Recording Code (ISRC) was developed to sit beside each digitally encoded audio track as a means of identification. It has many purposes, but the main one is for automation of royalty payments. Every time a song is played on the radio, systems are available to log that the track has been played and in turn, the appropriate royalty payment is transferred to the appropriate collection agencies.

Prior to the mastering session, it is advisable to ensure that all ISRC codes are obtained well in advance so that they can be encoded and added to the PQ sheet.

UNIVERSAL PRODUCT CODE/EUROPEAN ARTICLE NUMBER

The Universal Product Code (UPC) was originally introduced in the 1970s to log products as they left the grocery store in America. The EAN (European Article Number) is the European equivalent. The record label will normally supply this code. As with the ISRC, it is prudent to ensure that you have this code well in advance of the mastering session.

OTHER CODES

Other codes are often required at session along with other details such as any additional catalog or reference codes. Other information such as the record label or client will also be required at times.

Some of the main codes that are often asked about are things such as CD Text. CD Text can be programmed in a similar way to the user programming information about the track into a Minidisc machine. CD Text will show up on many CD players and is occasionally used, although it is less common these days.

DELIVERY

To the Mastering House

For the delivery of audio material, mastering houses are ever more commonly working remotely from the production team, receiving files over the Internet using File Transfer Protocol (FTP). In the same manner, mastering engineers can also post their completed masters to the manufacturers using FTP too.

Lots of audio is still brought to the mastering house in the same way it always has been via analog tape, DAT, CDR and data. One thing that must be mentioned at this stage is labeling. Make sure that your labeling is something you understand and can be conveyed easily to the mastering house. Which file is the master? Which is the copy? Which version are we supposed to be using? Clarity on such issues will save time and lower costs in the long run.

Mastering engineers will of course vary but most will prefer you to send them the highest resolution file you can. Certainly, 24 bit and as high a sample rate as the recording was made on. Typically, this is 44.1 kHz and 24 bit. However, higher sample rates are often used and should be passed to the mastering house. Leave it to the mastering engineer to get it to 16 bit 44.1 kHz CD-ready audio.

To the Manufacturer

Mastering houses will produce the manufacture-ready master in a range of formats. Most typically this will be a Disc Description Protocol image file (DDPi).

```
///////////////////////////////////////////////////////////////////////////
//
//    Table of Content generated by Pyramix Virtual Studio CD Mastering System
//
///////////////////////////////////////////////////////////////////////////


Disc Title        : Guitar Revival 2
Label             : Universal Publishing Production Music
Date              : 15 March 2008

Customer Name     : Universal Publishing Production Music
Customer Contact  : Joe Bloggs
Customer Phone    : 020712345678

Master ID Code    :
Ref. Code         : C07-20
UPCEAN Code       : 0000000000000

Track #    Index #    Time         ISRC/Name            Copy

01         00         00:00:00     GBAZC0834301   yes
           01         00:02:00     Sabotage
Length                02:44:56
-------------------------------------------------------------
02         00         02:46:56     GBAZC0834302   yes
           01         02:48:34     East Bay Livin'
Length                01:56:74
-------------------------------------------------------------
03         00         04:45:33     GBAZC0834303   yes
           01         04:46:64     Tight Jeans & Trilby Hats
Length                02:54:03
-------------------------------------------------------------
04         00         07:40:67     GBAZC0834304   yes
           01         07:42:71     Time to Shout
Length                02:41:43
-------------------------------------------------------------
05         00         10:24:39     GBAZC0834305   yes
           01         10:27:53     Arctic Melt
Length                02:56:07
-------------------------------------------------------------
06         00         13:23:60     GBAZC0834306   yes
           01         13:25:16     Jump Start
Length                02:33:05
-------------------------------------------------------------
07         00         15:58:21     GBAZC0834307   yes
           01         16:00:00     Lazy Eye
Length                03:24:17
-------------------------------------------------------------
08         00         19:24:17     GBAZC0834308   yes
           01         19:27:24     Take Me Out Tonight
Length                02:46:00
-------------------------------------------------------------
09         00         22:13:24     GBAZC0834309   yes
           01         22:14:36     Time On Our Side
Length                02:56:33
-------------------------------------------------------------
10         00         25:10:69     GBAZC0834310   yes
           01         25:12:56     Black Roses
Length                03:06:67
-------------------------------------------------------------
11         00         28:19:48     GBAZC0834311   yes
           01         28:21:27     The Night Is Right
Length                02:39:09
-------------------------------------------------------------
12         00         31:00:36     GBAZC0834312   yes
           01         31:01:71     Sidewalk
Length                02:55:48
-------------------------------------------------------------
13         00         33:57:44     GBAZC0834313   yes
           01         33:59:35     Arctic Melt (Acoustic)
Length                02:17:34
-------------------------------------------------------------
14         00         36:16:69     GBAZC0834314   yes
           01         36:19:21     Lazy Eye (Acoustic)
Length                03:28:30
-------------------------------------------------------------
15         00         39:47:51     GBAZC0834315   yes
           01         39:54:45     Time On Our Side (Acoustic)
Length                02:57:18
-------------------------------------------------------------
16         00         42:51:63     GBAZC0834316   yes
           01         42:55:10     Sidewalk (Acoustic)
Length                02:47:05
-------------------------------------------------------------
17         00         45:42:15     GBAZC0834317   yes


01         45:43:73     Sabotage 30
Length                00:30:00
-------------------------------------------------------------
18         00         46:13:73     GBAZC0834318   yes
           01         46:15:55     East Bay Livin' 30
Length                00:30:00
-------------------------------------------------------------
19         00         46:45:55     GBAZC0834319   yes
           01         46:47:37     Tight Jeans & Trilby Hats 30
Length                00:30:00
-------------------------------------------------------------
20         00         47:17:37     GBAZC0834320   yes
           01         47:19:19     Time To Shout 30
Length                00:30:00
-------------------------------------------------------------
21         00         47:49:19     GBAZC0834321   yes
           01         47:51:01     Arctic Melt 30
Length                00:30:00
-------------------------------------------------------------
22         00         48:21:01     GBAZC0834322   yes
           01         48:22:58     Jump Start 30
Length                00:30:00
-------------------------------------------------------------
23         00         48:52:58     GBAZC0834323   yes
           01         48:54:41     Lazy Eye 30
Length                00:30:00
-------------------------------------------------------------
24         00         49:24:41     GBAZC0834324   yes
           01         49:26:23     Take Me Out Tonight 30
Length                00:30:00
-------------------------------------------------------------
25         00         49:56:23     GBAZC0834325   yes
           01         49:58:05     Time On Our Side 30
Length                00:30:00
-------------------------------------------------------------
26         00         50:28:05     GBAZC0834326   yes
           01         50:29:62     Black Roses 30
Length                00:30:00
-------------------------------------------------------------
27         00         50:59:62     GBAZC0834327   yes
           01         51:01:44     The Night Is Right 30
Length                00:30:00
-------------------------------------------------------------
28         00         51:31:44     GBAZC0834328   yes
           01         51:33:26     Sidewalk 30
Length                00:30:00
-------------------------------------------------------------
29         00         52:03:26     GBAZC0834329   yes
           01         52:05:08     Arctic Melt (Acoustic) 30
Length                00:30:00
-------------------------------------------------------------
30         00         52:35:08     GBAZC0834330   yes
           01         52:36:65     Lazy Eye (Acoustic) 30
Length                00:30:00
-------------------------------------------------------------
31         00         53:06:65     GBAZC0834331   yes
           01         53:08:47     Time On Our Side (Acoustic) 30
Length                00:30:00
-------------------------------------------------------------
32         00         53:38:47     GBAZC0834332   yes
           01         53:40:29     Sidewalk (Acoustic) 30
Length                00:30:00
-------------------------------------------------------------
AA         01         54:10:29
```

FIGURE 6.1

Part of an example PQ Sheet showing all the relevant codes used such as ISRC.

These are a collection of data files that continues to be the most reliable. With a Disc Description Protocol (DDP), it will either work or it won't. DDP can either be stored on a CDR (if it will fit), a DVD-R, hard drive, exabyte tape, or uploaded to an FTP server. Manufacturers accept Audio CDs as the master for duplication, although this is not advisable given the opportunity for error on the disk. Ensure the PQ sheet with all the completed information tallies to the audio sent.

Some mastering engineers such as Ray Staff are producing enhanced CDs for their clients that include a data section. This data section holds some MP3s of the album's material that has been data compressed at the mastering house using specially prepared masters. These masters have been adjusted especially for the MP3 player market and with the perceptual coding systems used in mind, thus providing superior quality playback.

RAY STAFF ON FORMATS

"There have been many arguments over the years about the varying quality of manufactured CDs. I would suggest there are three main areas in which the quality can be influenced. The CD master, the glass mastering and the quality of the actual pressing.

There are now two main mastering formats available to us to send to the manufacturers. They are CD audio and DDP. The format used for your CD master is important for the quality of the final product and its storage as part of a record companies assets.

The CD is easy, convenient and supported by all good digital audio workstation (DAW) systems. It is essential to have a good quality CD writer running at the optimum speed (x1 is the speed of choice for many mastering engineers) and to have high quality CD blanks. Once written, the CD should be evaluated for disk errors using a CD analyzer and auditioned for content errors.

There is also some confusion about the terms PMCD (PreMaster CD) and Red Book CD. Red Book is the original specification laid down by Sony and Philips for the Audio CD (CDDA). PMCD is a format that was introduced by Sony, Sonic Solutions and Start Lab. There is a small additional piece of hidden data containing the "PQ or PreMaster Cue Sheet." Only Sonic Solutions systems can write a PMCD and only Laser Beam Recorders manufactured by Sony can read them. Essentially both offer the same quality.

I personally find the quality variations of this format unreliable. It takes a well-configured system with carefully selected blanks to produce good results. The quality of cloned copies is also very variable and I have often seen this type of CD master physically deteriorate reducing its reliability for long term storage and future use.

DDP—developed by Doug Carson & Associates—is a method of storing an audio CD as data. Only the more professional DAW systems include this option. DDP masters were originally written to an Exabyte tape. As a collection of files they can be stored on a hard drive (DDPi or DDP image file) and shipped via a ROM disk or via an ftp site. This format offers reduced delivery costs and speeds, high quality and efficient long-term storage. Many large record companies use DDP files to archive their product as they can be cloned repeatedly (using good error checking software) with no audible deterioration. Using large data archiving and media asset management systems they can have an infinite storage life and high access speeds for future manufacturing. Again Quality Control is essential and a CD should be made from the DDP master and auditioned for content errors before dispatch to the plant.

Most online music stores, will only accept an audio CD as the source for making the MP3's on their site. It would be a positive step forward if they too would accept a DDP master. With difficulty, we have sometimes persuaded some providers to use WAV files. The quality of many online or homemade lossy data compression files varies considerably. To this end, some mastering rooms do offer a service to make high quality files. On a number of projects we have been asked to add these to the data area of an enhanced CD to give the quality conscious consumer an alternative."

THE LOUDNESS WARS

Much of the mastering engineer's skill has been abused over recent years to develop what some in the industry have dubbed The Loudness Wars. Producers, engineers and mastering engineers all agree that levels have increased so much that they have literally hit the red!

As an example of the issue, I must *raise* the tone of this text to a scene from Spinal Tap. Nigel Tufnel (played by Christopher Guest) is showing "rockumentary" maker Marty DiBergi (Rob Reiner) his collection of guitars. Nigel proudly points out his custom guitar amp upon which the volume control markings go up to 11 and not the standard 10. Nigel proudly states that "when you get to 10, where can you go? Nowhere! ... this goes up to 11!" Rightly Marty says, well "why don't you just make 10 louder?" A dazed and confused Tufnel retorts after a long, painful and thought-ful pause "This goes one louder" (This is Spinal Tap, 1984). In all seriousness though we've come to a pivotal crossroads in the production of modern music that audio professionals are concerned about.

This Spinal Tap analogy actually works. The maximum loudness has been achieved by the system we have in place—digital audio. The only way we can go is down, as we cannot make "10 louder" without accepting a new digital standard. Digital audio will always have an upper limit of 0 dB in its current format and this is the level at which we're all aiming for and nothing less will seemingly do (see box "Ray Staff on Loudness"). However, at some point we have to say "no more," as audio quality is suffering overall despite some valiant efforts by the mastering community to make their masters still shine through these restrictions.

Recently, I have had the fortune to listen to the direct mixes of an album I really rate. The album was produced by an A-list producer and mixed by an exceptional team. The mixes were dynamic, wide, good use of frequency range and simply beautifully crafted. What happened at the mastering stage? The released album is distorted in some way at loud sections and does not translate very well on some excellent monitoring systems. Was the mastering engineer asked to raise the levels by inexperienced record label executives or the artist(s) themselves?

RAY STAFF ON LOUDNESS

"Digital recording and mastering levels have increased dramatically in recent years. Most digital equipment is supplied with digital peak reading meters only with a scale calibrated from 0 dB downward and a clip indicator. The manufacturers were keen that we should avoid the distortion caused by digital clipping. Unfortunately, the digital zero has become a target and people assume that if the meter is not hitting the highest possible level their recording will be deficient in some way and will not sound very loud.

The only meter in common use that gives any true indication of loudness is a Volume Unit (VU) meter. To maintain a consistent level across a number of songs they should all have a similar VU reading. The final relative level adjustments must be made by ear. Peaks within these tracks may vary considerably depending upon the musical content. When digital was first introduced most studios would set their analog console "0"VU to equal −14 on the digital scale leaving a respectable amount of headroom for transients. Many CDs are now so hot it has become necessary to calibrate "0"VU to equal −4 dB on the digital scale. That means if the average level is "0"VU there is only about 4 dB left for transients and peaks.

Some engineers at the remix stage are adding large amounts of compression and limiting just because they believe it is the right thing to do, with little consideration given to the resultant sound. All, too often, this is mixed within a computer that is struggling with the number crunching using cheap, poorly adjusted plug-ins. In the mastering room we are told "I think my mix sounds a bit small but we want our CD to sound big and loud." Unfortunately the damage has been done. Too much or the wrong type of dynamic processing can make music sound small with a lack of depth and punch. This is, of course, the worst possible scenario and fortunately there are many good mixes out there.

It is not wrong to use compression as it is an invaluable tool, but I really think that studios should check the effectiveness of any compression or limiting used in the mix. When using any audio processing that adds gain, it is important to compare it to the original with the level matched. Only then can you really determine if it is doing a worthwhile job. If in doubt don't do it. That gives the mastering engineer the opportunity to work with a recording that has got potential to be manipulated and worked into a great sounding CD. I am pleased that some DAW systems now offer the ability to use a VU scale alongside a peak reading meter. This now gives the operator the ability to have a better judgment of the loudness.

The mastering should be auditioned at different levels to ascertain the best possible result and consideration given to what happens after the mastering stage. Lossy media files such as MP3 may distort if the level is too hot and it is a fallacy that a CD has to be loud to sound good on the radio.

I am a keen supporter of the Bob Katz "K-System" (Katz, 2007). This is an extension of the original alignment method I described earlier and also links the recording level to an acoustic reference level. It's a good common sense approach to engineering and removes digital zero target."

What Is Loudness?

Loudness is one of those terms that are rather unquantifiable. Yes, we do have all manner of units of measurement for loudness in one form of energy or another. However, what they mean in each case can be interpreted in several ways. For example, in audio engineering we have several ways of explaining how loud something is and each has its own part in the engineering chain. Terms such as Gain, Level, Volume, Intensity, Power and Loudness all can mean the same thing but can be interpreted differently. These differences are not addressed here but outline some of the problems.

FIGURE 6.2
A Volume Unit (VU) Meter is perfect for showing the true loudness of a signal.

We measure how loud something is so that we can ensure that we capture the audio signal in the best possible way. So it is not distorted or too quiet, that it sits inside the analog noise floor, or is so low that it succumbs to quantization error (an artifact of many analog-to-digital conversion processors).

As Ray Staff has already pointed out, 0 dB on a modern peak meter has become a target as opposed to a warning. Peaks of music will be hitting this at all costs at the request of many clients. However, it is not the peaks that represent the loudness, but the RMS or average power, of the signal. This is best represented by what we call a VU meter (Figures 6.2 and 6.3).

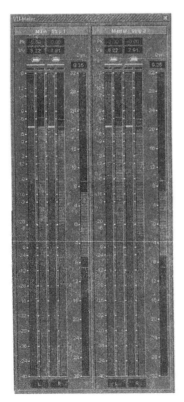

FIGURE 6.3
Merging's Pyramix stereo meters can show both peak meters (the louder of the ones shown) and VU meters (the lower), plus a useful Dynamics meter (to the very right) showing the difference between the peak and VU levels. In this example, the progam material is very loud showing an average level of around −6 dB below and peaking at 0 dBFS.

The problem of loudness in some modern records is exacerbated because each record is competing to be louder than each other. By making every record louder than the last means that the natural dynamics of the next audio material has to be less than the one before to fit into the set amplitude scale of digital audio.

THE K-SYSTEM

The management of dynamic range is an important one that Bob Katz addresses in his book entitled *Mastering Audio: The Art and the Science*. He outlines a metering system ("tied to monitoring gain" so that the perception of loudness is not altered) that proposes that the loudness of 0VU is regulated for the type of activity, whether that be cinema, music and so on. In each K-System, 0VU is intended to be linked to 83dBC SPL (for further details see Katz (2007)).

For most stages, such as recording and mixing, Katz proposes that this is set to −20 below full scale allowing plenty of headroom for the peaks K-20 as it is known in the K-System. Headroom is the space available in the amplitude range that allows for dynamic range. For mastering he proposes to use "K-14 for a calibrated mastering suite."

In such a proposal, 0VU would be the "target" as opposed to the full scale of the dynamic range. This would immediately improve the quality of the music at all points in the chain and at the same time allow for the natural dynamics that makes music "breathe."

The following chapters on mastering are by Bob Katz. Bob describes some of the key processes that will take place during the mastering session and the whys and what-fors of mastering with a pro. Reading these should prepare anyone for what they should do before the session and what to expect.

FURTHER READING

Cousins, M. & Hepworth-Sawyer, R. (2008). *Logic Pro 8: Audio & Music Production*. Focal Press, Oxford.
Huber, D. & Runstein, R. (2005). *Modern Recording Techniques*. Focal Press, Oxford.
Katz, B. (2007). *Mastering Audio: The Art and the Science*. Focal Press, Oxford.
www.mottosound.com.
www.digido.com.
www.ifpi.org/content/section_resources/isrc.html.

Chapter 6
B. Equalization Techniques

Bob Katz

In This Chapter

INTRODUCTION

The First Principle of Mastering

The first principle of mastering is this: changing anything affects everything. This principle means that mastering becomes the art of compromise, the art of knowing what is sonically possible, and then making informed decisions about what is most important for the music.

Equalization practice is an especially clear case of where a technique used in mastering is crucially different from an apparently similar technique used in mixing. For example, when mastering, adjusting the low bass of a stereo mix will affect the perception of the extreme highs. Similarly, if a snare drum sounds dull but the vocal sounds good, then many times the voice will suffer when you try to equalize for the snare. These problems occur even between elements in the same frequency range. During mixing, bass-range instruments that exhibit problems in their harmonic range can be treated individually, but in mastering their harmonic range overlaps with the range of other instruments. For example, although a mix engineer can significantly boost a bass instrument somewhere between 700 Hz and 2 kHz, mastering even a small boost in this range can have detrimental effects. Or when we need to fix a bass drum problem, to minimize affecting the bass guitar it may be necessary to try careful, selective equalization to "get under the bass" at the fundamental of the drum, somewhere under 60 Hz. Sometimes we can't tell if a problem can be solved until we try, so we try not to promise a client miracles!

WHAT IS A GOOD TONAL BALANCE?

Perhaps the major reason clients come to mastering houses is to verify and obtain a good tonal balance. But what, exactly, is a "good" tonal balance? The human ear responds positively to the tonality of a symphony orchestra which on a spectrum analyzer always shows a gradual high-frequency roll-off—as will most good pop music masters. The amount of this roll-off varies considerably depending on the musical style of course, so we use our ears, not the spectrum analyzer, as the basis for our EQ judgments.

Everything starts with the midrange: the fundamentals of the vocal, guitar, piano and other instruments must be correct, or nothing else can be made right. The message in the music—and more literally in radio, Internet and low-cost home systems—comes from the midrange. Listen to a great recording in the next room. The information still comes through despite the filtering of the doorway, carpets and obstacles. Then try filtering the recording severely below 200 Hz and above 5 kHz (like the sound of an old, bad cinema loudspeaker). A good recording will still translate.

The mastering engineer tries to make the sound pleasant, warm and clear, if that is correct for the genre. However, a master can deviate from this to provide a deliberately different color—for example, a brighter, thinner sound,[1] the mastering engineer controls excessive deviation from neutral ensuring that the sound will translate to the widest variety of playback systems and on the air.[2]

---

[1] We may believe we have "the absolute sound" in our heads but are surprised to learn how much the ear/brain accommodates. If we play a bright album immediately after a dark one, at first there is an ear shock, but we quickly adapt, though the new sound continues to affect subliminally. The same thing happens in photography and motion pictures, after an initial shock, the "Kodachrome effect" becomes subliminal.

[2] Overly bright records can become dull on the air due to high frequency FM broadcast limiters; radio processing makes brightness self-defeating.

Specialized Music Genres

The symphonic tonal balance is a generally good guide for rock, pop, jazz, world music and folk music, especially in the mid to high frequencies. But some specialized music genres deliberately utilize very different frequency balances. We could think of reggae as the symphony spectrum with lot more bass instruments, whereas punk rock is often extremely aggressive, thin, loud and bright. Punk voices can be thin and tinny over a fat musical background. If this straining of the natural fundamental–harmonic relationships is excessive and done for a whole record most people would find it fatiguing, but it can be interesting when it's part of the artistic variety of the record.

Be Aware of the Intentions of the Mix

Equalization affects more than just tonality—it can also affect the internal balance of a mix. So a good mastering engineer must make sure he/she understands the intentions and vision of the production team. In fact, mastering equalization may help the producer's balance if his/her judgment had been inadvertently affected by a monitoring problem in the mix environment.

EQUALIZATION TECHNIQUES

Two Basic Types: Parametric and Shelving

There are two basic types of equalizers—parametric and shelving—named after the shape of their characteristic curve. Parametric EQ, invented by George Massenburg circa 1967,[3] is the most flexible curve, providing three controls: center frequency, bandwidth and level of boost or cut. Mix engineers like to use parametrics on individual instruments, either boosting to bring out their clarity or salient characteristics, or selectively dipping to eliminate problems, or by virtue of the dip, to exaggerate the other ranges. The parametric is also the most popular equalizer in mastering as it can be used surgically to remove certain defects, such as overly resonant bass instruments or enhance narrow ranges of frequencies. However, shelving equalizers are more popular in mastering than in mixing, as they provide boosts or cuts to the entire spectrum below or above a selected frequency, and can alter the tonality of the entire mix.

Parametric: Q and Bandwidth

The parameter Q is defined mathematically as the product of the center frequency divided by the bandwidth in Hertz at the 3 dB down (up) points measured from the peak (dip) of the curve. A low Q means a high bandwidth and vice versa.

---

[3]In 1967, George Massenburg began the search for a circuit that would be able to independently adjust an equalizer's gain, bandwidth and frequency. The key word is independent, for most analog circuits fail in this regard and the frequency, Q, and gain controls interact with each other. His circuit, which he dubbed a parametric equalizer, remains proprietary today.

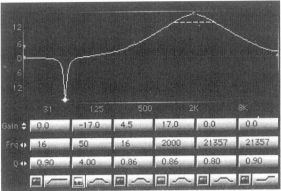

FIGURE 6.4
Parametric equalizer
with +17 dB boost
centered at 2 kHz
with a fairly wide
bandwidth of 1.60
octaves (Q = 0.86),
indicated by the
dashed white line at
the 3 dB down points.
A cut of −17 dB at
50 Hz with a very
narrow bandwidth of
0.36 octaves (Q = 4).

Figure 6.4 shows two parametric bands with extreme levels for purposes of illustration. On the left, a 17 dB cut at 50 Hz with a very narrow Q of 4, which is 0.36 octaves. The bandwidth is 12.5 Hz. On the right, a 17 dB boost centered at 2 kHz, with a fairly wide (gentle) Q of 0.86, which is 1.6 octaves. The bandwidth is 2325 Hz, represented by the dashed white line.[4]

The choice of high or low Q depends on the situation. Gentle equalizer slopes almost always sound more natural than sharp ones, so Qs of 0.6 and 0.7 are therefore very popular. Higher (sharper) Qs (greater than 2) are used surgically, to deal with narrow-band resonances or discrete-frequency noises, though we must listen for artifacts of high Q such as ringing. It is possible to work on just one note with a sufficiently narrow-band equalizer, or we may overturn the first principle of mastering by using a higher Q to attempt to isolate and emphasize a single instrument. For example, a poorly mixed program may have a weak bass instrument; but boosting the bass around 80 Hz may help the bass but muddy the vocal, so we narrow the bandwidth of the bass boost. Rarely is this totally effective, so if the bass boost is not good for the vocal it's probably not good for the song. But if the vocal is made only slightly bassier, we may try a slight compensatory boost around 5 kHz, as long as that doesn't interact poorly with yet another instrument!

Focusing the Equalizer

There are at least two techniques for identifying a problem frequency. The classic approach is to focus the equalizer directly: starting with a large boost and fairly wide (low value) Q, sweep through the frequencies until the resonance is most exaggerated, then narrow the Q to be surgical, and finally, dip the EQ the amount desired. This technique works well with analog equalizers, but some digital equalizers present ergonomic obstacles: the inefficient mouse and latency. A second technique is for engineers who have a musical background—keep a keyboard handy to determine the key of the song and use your sense of relative pitch to determine the problem note. Then translate that note to frequency with a converter and dip that frequency. Pictured is an example of a bass EQ that I found in seconds using this method. It combines a shelving boost with a single dip at the problem frequency. Engineers with a Crane Song Ibis EQ can skip the converter, because this equalizer is marked directly in musical notes (Figure 6.5).

Shelving Equalizers

As I mentioned, a shelving equalizer affects the level of the entire low-frequency or high-frequency range below or above a specified frequency. Some have Q controls, defined as the slope of the shelf at its 3 dB up or down point. One interesting variant on the standard shelf shape can be found in the Weiss EQ-1, Waves Renaissance EQ

---

[4]Some equalizers define bandwidth in octaves instead of Q.

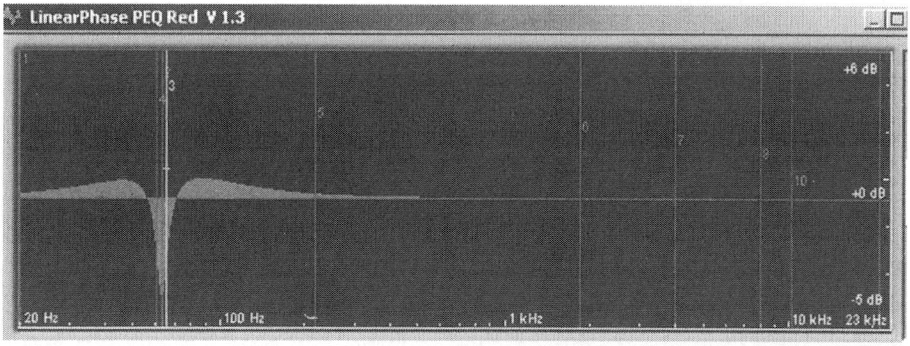

FIGURE 6.5
"One note bass" resonance fixed by a combination of a shelving boost (which was useful to help the rest of the notes that were weak) and a narrow-band dip at the resonance frequency.

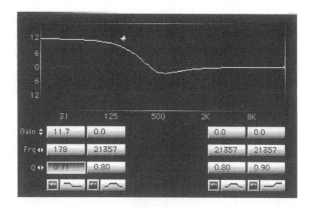

FIGURE 6.6
Gerzon resonant shelf with a low Q.

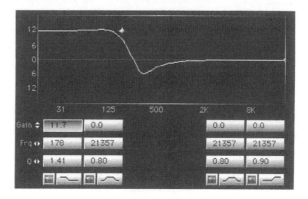

FIGURE 6.7
The dip just past the shelving boost frequency is characteristic of the Gerzon resonant shelf with a high Q.

and Manley's Massive Passive. This shape, called a resonant shelf, was proposed by psychoacoustician Michael Gerzon. I like to think of it as a combination of a shelving boost and a parametric dip (or vice versa). In Figure 6.6 a low Q (0.71) bass shelf of 11.7 dB below 178 Hz is mollified by a gentle parametric dip above 178 Hz; this resonant shelf is controlled by a single band of the equalizer. This type of curve can help keep a vocal from sounding thick while implementing a bass boost. Figure 6.7 shows the same boost with a high Q of 1.41.

EQ Yin and Yang

Remember the yin and the yang: contrasting ranges have an interactive effect.

For example:

- Adding low frequencies makes the highs seem duller and reducing them makes the sound seem brighter.
- Adding extreme highs between 15 and 20 kHz makes the sound seem thinner in the bass/lower midrange and vice versa.
- A slight dip in the lower midrange (around 250 Hz) reduces warmth and has a similar effect to a boost in the presence range (around 5 kHz).
- A harsh-sounding trumpet-section can be improved by dipping around 6–8 kHz and/or by boosting around 250 Hz.

Yin and yang considerations allow us to work in either or both contrasting ranges to ensure that the sound is both warm and clear. Thus, when the overall level is too high—we pick the range that we would be reducing. When an instrument exhibits upper midrange harshness—we pick the frequency range that would have the least effect on other instruments that are playing at the same time.

Using Baxandall for Air

In audio engineering, the air band is the range of frequencies about 15–20 kHz: the highest frequencies we can hear. An accurate monitoring system will indicate whether these frequencies need help. An air boost is contraindicated if it makes the sound harsh or unintentionally brings instruments like the cymbals forward in the depth picture. A third and important shape that's extremely useful in mastering is the Baxandall curve (right hand side of Figure 6.8), named after Peter Baxandall. Hi-Fi tone controls are usually modeled around this curve. Like shelving equalizers, a Baxandall curve is applied to low- or high-frequency boost/cuts. Instead of reaching a plateau (shelf), the

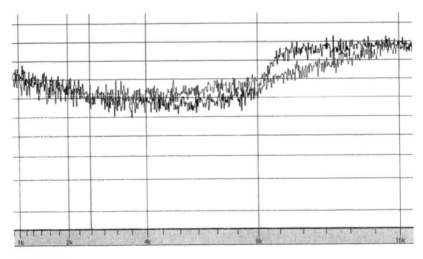

FIGURE 6.8
Gentle Baxandall curve (light) vs. sharp Q shelf (dark). Many shelving equalizers have gentler curves and may approach the shape of the Baxandall.

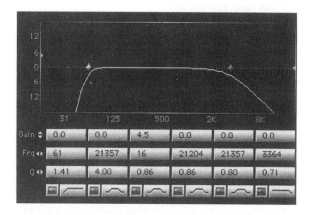

FIGURE 6.9
Sharp high-pass filter
at 61 Hz (at left) and
gentle low-pass filter
at 3364 Hz (at right).

Baxandall continues to rise (or dip, if cutting instead of boosting). Think of the spread wings of a butterfly with a gentle curve applied. This shape often corresponds better to the ear's desires than any standard shelf and a Baxandall high-frequency boost makes a great air EQ. Shelving equalizers with slope or Q controls can often approximate a Baxandall shape. Or, you can simulate a Baxandall high-frequency boost by placing a parametric equalizer (Q = ~1) at the high-frequency limit (~20 kHz). The portion of the bell curve above 20 kHz is ignored, and the result is a gradual rise starting at about 10 kHz and reaching its extreme at 20 kHz.

Care should be taken when making high-frequency boosts (adding *sparklies*). They are initially seductive but can easily become fatiguing. And the principle of yin and yang reminds us that the ear interprets a high-frequency boost as a thinning of the lower midrange. In addition, when the highs come up, the cymbals, triangle and tambourine become louder, which changes the balance of rhythm to melody, for better or worse.

High- and Low-Pass Filters

On the left-hand side of the EQ curve in Figure 6.9 is a sharp high-pass (low cut) filter at 61 Hz, and on the right-hand side, a gentle low-pass (high cut) filter at 3364 Hz. The frequencies are defined as the points where the filter is 3 dB down. Although they can be used to solve noise problems in mastering, they can also introduce problems of their own and they cannot be very surgical because they affect everything above or below a certain frequency. High-pass filters can reduce rumble, thumps, p-pops and similar noises. Low-pass filters are sometimes used to reduce hiss, since the ear is most sensitive to hiss in 3 kHz range, a parametric dip around that frequency is more effective than the radical low-pass filter. For hiss removal, we usually prefer specialized noise-reduction solutions over static filters (see Chapter 9).

One Channel or Both (All)?

Most times making the same EQ adjustment in both (all) channels is the best way to proceed as it maintains the stereo (surround) balance and the relative phase between channels. But sometimes it is essential to be able to alter only one channel's EQ. For example, with a too-bright high hat on the right side, a good vocal in the middle and proper crash cymbal on the left, the best solution is to work on the right channel's high frequencies.

Start Subtly First

Sometimes important instruments need help, though, ideally, they should have been fixed in the mix. The best repair approach is to start subtly and advance to severity only if subtlety doesn't work. For example, if the piano solo is weak, we try to make the changes surgically:

- only during the solo;
- only on the channel where the piano is primarily located, if that sounds less obtrusive;
- only in the frequency range(s) that help: fundamental, harmonic, or both;
- as a last resort by raising the entire level, because it would affect the entire mix, though the ear focuses on the primary instrument.

The Limitations and the Potential of the Recording

Waiting until the mastering stage to fix certain problems is inviting a compromise as there is only so much that can be accomplished in mastering. There is little we can do to fix a recording where one instrument or voice requires one type of equalization and the rest requires another.

Comb filtering is a similar issue. Equalization can do little or nothing to fix a comb filtering problem since EQ affects the entire mix. First we discuss the problem with the mix engineer to see if he/she can address the offending track and remix. If that is not possible, then we may try an overall mastering EQ, for example, a lower mid-range EQ boost to help a vocal that sounds thin due to comb filtering.

The better the mix we get, the better the master we can make, which implies, of course, that a perfect mix needs no mastering at all! And this is true: we should not automatically equalize; we should always listen and evaluate first. Many pieces leave mastering with no equalization or processing.

Instant A/Bs?

With a good monitoring setup, equalization changes of less than ½ dB are audible. I take an equalizer in and out to confirm initial settings, but then I avoid making instant EQ judgments because quickly switching back and forth distracts us from the long-term effect of an EQ change. Music is fluid so changes in the music can easily be confused with EQ changes. I usually play a passage for a reasonably long time with setting "A," then play it again with setting "B."

Loud and Soft Passages

I usually begin mastering on the loudest part of a song. Why? Because psycho-acousticians note that EQ peaks affect partial loudness more than dips, and loud passages accentuate these peaks more than soft ones. Equalization choices that are pleasing during a mezzo-forte passage may well displease during a loud one.

Fundamental or Harmonic?

The extreme treble range mostly contains instrumental harmonics. As the fundamental of a crash cymbal can be lower than 1.5 kHz, boosting the harmonics too much makes a cymbal sound tinny or thin. When equalizing or processing bass frequencies, it is easy to confuse the fundamental with the second harmonic.

You can note the parallel run of the bass instrument's fundamental from 62 to 125 Hz and its second and third harmonics from 125 to 250 Hz and up. Should we equalize the bass instrument's fundamental or the harmonic? It's easy to be fooled by the octave relationship; the answer has to be determined by ear—sometimes one, the other or both.

Bass Boosts Can Create Serious Problems

As the ear is significantly less sensitive to bass energy, bass information uses a lot more power for equal sonic impact: around 6–10 dB more below about 50 Hz, and about 3–5 dB more between 50 and 100 Hz.[5] This explains why bass instruments often have to be compressed to sound even. It also means that a low-frequency boost introduces so much energy it can reduce the highest clean intrinsic level we can give to a song (in cases where the client is demanding a "loud" master). Fortunately, the ear's tendency to supply missing fundamentals works in our favor, allowing us to save "energy" by cutting with a fairly sharp high-pass filter, but only if it does not hurt the energy of the bass drum or the low notes of the bass. The high-pass filter must be extremely transparent and have low distortion. Sometimes a gentle filter is a better choice than a steep one, as when dealing with a boomy bass drum or bass. But subsonic rumble or thumps benefit from a steep filter to have minimal effect on the instruments.[6]

Mix engineers working with limited bandwidth monitors run the risk of producing an inferior product. Utilizing good subwoofers permits you to hear low-frequency leakage problems that tend to muddy up the mix, for example, bass drum leakage in vocal and piano mikes. It's much better to apply selective high-pass filtering during the mixing process because mastering filters will affect all the instruments in a frequency range. For example, mix engineers may get away with a steep 80 Hz filter on an isolated vocal, but that's generally too high a frequency for mastering a whole mix. A mixing engineer should form an alliance with a mastering engineer, who can review his/her first mix and alert his/her to potential problems before they get to the mastering stage.

OTHER REFINEMENTS

Linear Phase Equalizers: The Theory

All current analog equalizer designs and nearly all current digital equalizers produce phase shift when boosted or cut; that is, signal delay varies with frequency and the length of the delay changes with the amount of boost or cut. The higher the Q, the more the phase shift. This kind of filter will always alter the musical

---

[5]This is dictated by the psychoacoustic equal loudness curves, first researched by Fletcher, Harvey and Munson in the 1930s.
[6]Historically, the high-pass filter was crucial when making LPs, to prevent excess groove excursion and obtain more time per LP side, but digital media do not have this physical problem.

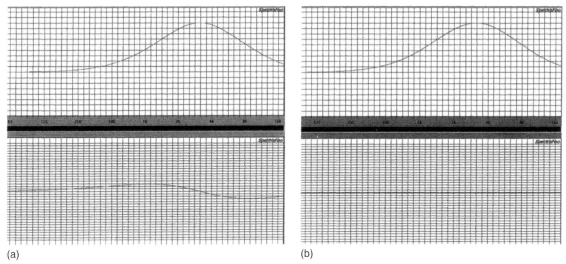

(a)   (b)

FIGURE 6.10
A minimum phase EQ boost exhibits phase shift (a), but a linear phase EQ boost has none (b).

timing and wave shape, also known as phase distortion. Daniel Weiss introduces:

"(In contrast) a particular type of digital filter, called the **Symmetric FIR Filter**, is inherently linear phase.[7] This means that the delay induced by processing is constant across the whole spectrum, unconstrained by EQ settings."[8]

In Figure 6.10 we graph the amplitude and phase response of a minimum phase EQ contrasted with a linear phase. The linear phase EQ phase response is a straight line.

Linear Phase EQ: The Sound

To my ears, the linear phase (abbreviated LP here) sounds more analog like than even analog, but ironically, while mastering a punk rock recording, my EQ proved too sweet in LP mode so I had to return to normal mode to give the sound some grunge. Clearly much of the qualities we've grown accustomed to in standard equalizers (also known as minimum phase equalizers, abbreviated MP) must be due to their phase shift; in fact John Watkinson believes that much of the audible difference between EQs comes down to their different phase response.[9] The Weiss has a very pure tone quality and seems to boost and cut frequencies without introducing obvious artifacts.

Neither approach is fundamentally better. It's easier to be aggressive with MP. Because of the pre-echo alluded to by Jim Johnston in my book, the LP subtly reduces transient response, which might contribute to its "sweeter," softer sound. The MP's tendency to smear depth can yield a pleasant artificial depth where

[7]FIR means Finite Impulse Response, and IIR means Infinite Impulse Response. Read John Watkinson's "The Art of Digital Audio" for an explanation of the differences.
[8]Weiss, Daniel, Weiss website, check www.digido.com/links/ for the Weiss website.
[9](9/1997) Studio Sound Magazine.

©2009 Bob Katz

none existed before. However, the LP's ability to preserve the depth of the mix is frequently an advantage when mastering, because altering a band's level does not move instruments forward or backward in the soundstage, as would happen with a MP equalizer. Narrow-band peaks and dips can be accomplished with no artifacts in LP, avoiding the smeary quality that occurs in MP. I especially love the ability of LP to keep high hats from moving forward when boosting in the cymbal range. Alan Silverman (in correspondence) says:

> "EQ'ing frequency ranges (in LP) is more like raising or lowering a fader in a mix than EQ'ing."

Dynamic Equalization

Dynamic equalizers (like the Weiss EQ1-Dyn) emphasize or cut frequency ranges dynamically, as opposed to static or fixed EQ. They have threshold controls which set a level above or below which EQ is dynamically boosted or cut, adding this extra boost or cut to a static setting. For example, above the threshold we can lower the high-frequency response; we could start a static band at say +1 dB, but to prevent harshness at high levels slowly cut the band's level when the signal exceeds the threshold. Dynamic EQs can be used as noise or hiss gates, rumble filters that function at low levels (especially useful for traffic control in a delicate classical piece), sibilance controllers or ambience enhancers. They can enhance inner details or clarity of high frequencies at low levels, where details are often masked by noise. Or enhance warmth by raising the respective band at low levels but prevent the sound from getting muddy at high levels. Multiband dynamics processors discussed later can also perform dynamic equalization.

C. Dynamic Range 1: Macrodynamics

Bob Katz

In This Chapter

Myth: "Of course I've got dynamic range. I'm playing as loudly as I can!"[10]

THE ART OF DYNAMIC RANGE

The term dynamic range refers to the difference between the loudest and softest passages of the body of the music; so it should not be confused with loudness or absolute level. The dynamic range of popular music is typically only 6 to 10 dB but for some musical forms it can be as little as 1 dB or as great as 15 dB (very rare). In typical pop music, soft passages 8 to 15 dB below the highest level are effective only for brief periods but in classical, jazz and many other acoustic forms, soft passages can effectively last much longer.

---

[10]A common misconception. Thanks to Gordon Reid of Cedar for contributing this audio myth.

Microdynamics and Macrodynamics

Dynamics can be divided further into Microdynamics—the music's rhythmic expression, integrity or bounce; and Macrodynamics—the loudness differences between sections of a song or song cycle. Usually dynamics processors (such as compressors, expanders) are best for microdynamic manipulation, and manual gain riding is best for macrodynamic manipulation. The microdynamic and macrodynamic work hand in hand, and many good compositions incorporate both microdynamic changes (e.g., percussive hits or instantaneous changes) and macrodynamic (e.g., crescendos and decrescendos). If you think of a music album as a four-course meal, then the progression from soup to appetizer to main course and dessert is the macrodynamics. The spicy impact of each morsel is the microdynamics. In this chapter we concentrate on macrodynamics.

The Art of Decreasing Dynamic Range

The dynamics of a song or song cycle are critical to creative musicians and composers. As a mastering engineer, my paradigm sound quality reference should be a live performance; I should be able to tell by listening if a recording will be helped or hurt by modifying its dynamics. In natural performance, the choruses should sound louder than the verses, and the climax should be the loudest. Many recordings have already gone through several stages of transient-destroying degradation, and indiscriminate or further dynamic reduction can easily take the clarity and the quality downhill. However, usually the recording medium and intended listening environment simply cannot keep up with the full dynamic range of real life, so the mastering engineer is often called upon to raise the level of soft passages, and/or to reduce loud passages, which is a form of manual compression.[11] Engineers may reduce dynamic range (compress) when the original range is too large for the typical home environment, or to help make the mix sound more exciting, fatter, more coherent, to bring out inner details, or to even out dynamic changes within a song if they sound excessive.[12]

Experience tells the mastering engineer when a passage is too soft. The context of the soft passage also determines whether it has to be raised. For example, a soft introduction immediately after a loud song may have to be raised, but a similar soft passage in the middle of a piece may be just fine. This is because the ears self-adjust their sensitivity over a medium time period and may not be prepared for an instantaneous soft level after a very loud one. Thus, meter readings are fairly

---

[11]Please do not confuse the term dynamic range reduction (compression) with data rate reduction. Digital coding systems employ data rate reduction, so that the bit rate (measured in kilobits per second) is less. Examples include the MPEG (MP3) or Dolby AC-3 (now called Dolby Digital) systems. Since it's not good to refer to two different concepts with the same word, we should encourage people to use the term data reduction system or coding system when referring to data and compression only when referring to the reduction of dynamic range.

[12]Excessive is definitely a subjective judgment! It's very important to develop an esthetic that appreciates the benefits of dynamic range, and which also knows when there is too much—or too little. This is both a matter of taste, as well as objective knowledge of the requirements of the medium and listening environment.

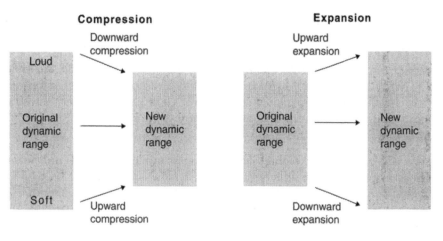

FIGURE 6.11
Any combination of these four processes may be employed in a mastering session.

useless in this regard. How soft is too soft? The engineers at Lucasfilm discovered that having a calibrated monitor gain and a dubbing stage with NC-20[13] noise floor do not guarantee that a film mix will translate to the theater. During theater test screenings, some very delicate dialog scenes were "eaten up" by the air conditioning rumble and audience noise in a real theater. So they created a calibrated noise generator, labeled "popcorn noise," which could be turned on and added to the monitor mix whenever they wanted to check a particularly soft passage. For similar purposes, our alternate listening room here at Digital Domain has a ceiling fan and other noisemakers. Whenever I have a concern, I start the digital audio workstation (DAW) playing a loud passage just before the soft one, and take a walk to the noisy listening room.

The Art of Increasing Dynamic Range

This can also make a song sound more exciting, by using contrast or by increasing the intensity of a peak. The key to success here is to recognize when an enhancement has become a defect—musical interest can be enhanced by variety, but too much variety is just as bad as too much similarity. Another reason to increase dynamic range is to restore or attempt to restore the excitement of dynamics that had been lost due to multiple generations of compression or tape saturation.

The Four Varieties of Dynamic Range Modification

We always use the term compression for the reduction of dynamic range and expansion for its increase. There are two varieties of each: upward compression, downward compression, upward expansion and downward expansion, as illustrated in Figure 6.11.

Downward compression is the most popular form of dynamic modification, bringing high-level passages down. Limiting is a special case—downward compression with a very high ratio is explained later in this chapter. Examples include just about every compressor or limiter you have ever used.

---

[13]A room with NC-30 rating is quiet. The quietest rooms may have noise floors lower than NC-20.

Upward compression raises the level of low passages. Examples include the encode side of a Dolby® or other noise reduction system, the automatic gain control (AGC) which radio stations use to make soft things louder, and the type of compressor frequently used in consumer video cameras.[14] In my book *Mastering Audio: The Art and the Science* I explain a more effective upward compression technique that is extremely transparent to the ear. For clarity, we will always use the short-term compressor to mean downward compressor unless we need to distinguish it from upward compressor.

Upward expansion takes high-level passages and brings them up even further. Upward expanders are still rare; in skilled hands they can be used to enhance dynamics, increase musical excitement or restore lost dynamics. Examples include the peak restoration process in the playback side of a Dolby SR, the DBX Quantum Processor, the various Waves brand dynamics processors and the Weiss DS1-MK2 when used with ratios less than 1:1 (to be explained).

Downward expansion is the most common type of expansion: it brings low-level passages down further. Most downward expanders are used to reduce noise, hiss or leakage. A dedicated noise gate is a special case—downward expansion with a very high ratio (to be explained). Examples of downward expanders include the classic Kepex and Drawmer gates, Dolby and similar noise reduction systems in playback mode, expander functions in multi-function boxes (e.g., TC Finalizer) and the gates on recording consoles. Again for clarity, we will use the simple term expander to mean the downward type unless we need to distinguish it from the upward type.

THE ART OF MANUAL GAIN RIDING: MACRODYNAMIC MANIPULATION

In General

During mixing it's difficult to simultaneously pay attention to the internal balances as well as the dynamic movement of the music between, for example, verse and chorus. Sometimes engineers inadvertently lower the master fader during the mix to keep it from overloading, which strips the climax of its impact. In mastering, we can enhance a well-balanced rock or pop mix by taking the dynamic movement of the music where it would like to go. Delicate-level changes can make a big difference; it's amazing what a single dB can accomplish. It's also our responsibility to make sure the client-level change was not intentional!

How and When to Move the Fader

Artistic-level changes can really improve a production but they need to be made in the most musical way. To this end, internal-level changes are least intrusive when performed manually (by raising or lowering the fader), as little as ¼dB at a time, as opposed to using processors such as compressors or expanders, which tend to be more aggressive.

---

[14]AGC has been given a bad name by a poorly implemented use in camcorders, which often exhibit pumping and breathing artifacts.

When riding the gain, aim just to augment the natural dynamic flow: if the musicians are trying for upward impact, pulling the fader back during a crescendo can be detrimental as it will diminish the intended impact. Extra-soft beginnings, endings or even middle spots require special attention. If the highest point in the song sounds "just right" after processing but the intro sounds too soft, it's best to simply raise the intro, finding just the right editing method to restore the gain to normal afterward using one or more of these approaches:

- A long, gradual lowering of the gain, which might occur at the end of the intro, or slowly during the first verse of the body.
- A series of ¼ or ½ dB edits, taking the sound down step by step at critical moments. This is useful when you don't want the listener to note that you're cheating the gain back down and you may be forced to work against the natural dynamics.
- A quick edit and level change at the transition between the raised-level intro and the normal-level body. This can have a nice effect and be the least intrusive.

The Art of Changing Internal Levels of a Song

Some soft passages must be raised, but if the musicians are trying to play something delicately, pushing the fader too far can ruin the effect. The art is to know how far to raise it without losing the feeling of being soft and the ideal speed to move the fader without being noticed. In a DAW, physical fader moves can be replaced by recording the movement of a fader on screen using a mouse or control surface, or by the use of cross-fades or by drawing-level changes on an automation curve. The mastering engineer's aim is to be invisible; if the sound is being audibly manipulated, the job has not been done properly.[15] Here's a technique for decreasing dynamic range in the least damaging way that I learned many years ago from Alec Nesbitt's book *The Sound Studio* (see Further Reading). If you have to take a loud passage down then the best place to lower the gain is at the end of the preceding soft passage before the loud part begins. Look for a natural dip or decrease in energy, and apply the gain drop during the end of the soft passage before the crescendo into the loud part begins. That way, the loud passage will not lose its comparative impact, for the ear judges loud passages in the context of the soft ones.

Figure 6.12 from a Sonic Solutions workstation illustrates the technique. The gain change is accomplished through a cross-fade from one gain to another.

The producer and I decided that the shout chorus of this jazz piece was a bit overplayed and had to be brought down from triple to double forte (which amounted to 1 dB or so).[16] To retain impact of the chorus, we slowly dropped the level during the soft passage just before the drum hit announcing the chorus. In the trim

---

[15]This is true for most of the "natural" music genres, with some exceptions being hip-hop, psychedelic rock, performance art, and so on, where the artists invite the engineer to contribute surprising or rococo dynamic effects.

[16]Producers don't always use classical Italian dynamic terms to describe their needs. The mastering engineer should choose the bonding language which is best for the client—"Make it louder, man!"

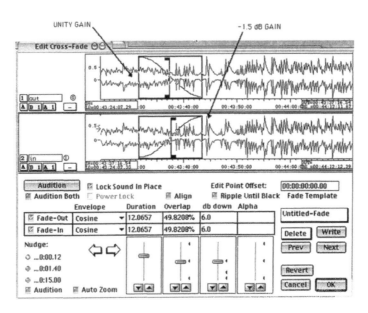

FIGURE 6.12
The modern version of fader riding. In Sonic Solutions' edit window the outgoing edit is on top, the incoming on the bottom. Note that the gain drop is performed in the soft passage preceding the loud downbeat, thus preserving the apparent impact of the downbeat.

window, we constructed a 12 second cross-fade from unity gain (top panel) to −1.5 dB gain (bottom panel); the drum hit is just to the right of the cross-fade box. This drum hit retains its impact by contrast, because the musicians' prior delicate decrescendo has been enhanced during mastering.

Where I Begin

I like to start mastering by going directly to the loudest part of the loudest song. This sets my monitor level and degree of processing for the album and reduces the later temptation to raise the loud part so much that excessive processing might squash it. After I get a great sound with the necessary processing, I return to the beginning of the song to see how it sounds in that context.

Having heard that, I may decide to reduce the level of a song's introduction. In Figure 6.13, I reduced the intro and slowly introduced a crescendo (20 seconds long) that enhances the natural build as it goes into the first chorus. The top panel is at –1 dB gain, bottom panel is at unity (0 dB) gain, achieved at the end of the cross-fade.

FIGURE 6.13
A soft introduction has been reduced even further, and the impact of the body of the song is enhanced by gradually increasing the gain during the beginning of the main part of the song.

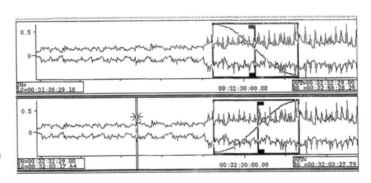

Increasing the space between two songs is another way of increasing the dynamic impact as it extends the tension caused by silence.

CONCLUSION

Macrodynamic manipulation is a sometimes overlooked but powerful tool in the mastering engineer's arsenal. In the next section we move on to the use of compressors, expanders and limiters to manipulate microdynamics.

FURTHER READING

This is an edited chapter from Katz, B. (2007). *Mastering Audio: The Art and the Science*. Focal Press, Oxford.

D. Dynamic Range 2: Dynamic Manipulation

Bob Katz

In This Chapter

157

In this part, Bob Katz speaks of the all-important dynamics processing approaches that the mastering engineer can apply to your mixes to get your masters sounding great.

COMPRESSORS AND LIMITERS: OBJECTIVE CHARACTERISTICS

This part of the chapter is about micro- (and some macro-) dynamic manipulation, which is achieved primarily through the use of dedicated dynamics processors. Here we look in detail at how compressors and limiters work, because we must first study their objective characteristics to understand how they can be successfully deployed on our music.

Transfer Curves (Compressors and Limiters)

A transfer curve is a picture of the input-to-output gain characteristic of an amplifier or processor. The transfer curve of a unity-gain amplifier shows a straight diagonal line across the middle at 45° called the unity-gain line. Unity gain means the ratio of output to input level is 1 or 0 dB. Figure 6.14 shows a family of linear curves.

Input level is plotted on the X-axis and output on the Y-axis. From left to right: unity gain, 10 dB gain, 10 dB attenuation. A linear amplifier is one which shows a straight line (not a curve) at 45°, hence the name. Notice that the middle plot would yield distortion for any input signals above −10 dBFS.

The threshold of a compressor is the level at which gain reduction begins, and compression ratio describes the relation between the input and output above the threshold. Figure 6.15a shows a simple compressor with a fairly gentle 2.5:1 compression ratio and a threshold at around −40 dBFS (which is quite low and would yield strong compression for loud signals). 2.5:1 means that an increase in the input of 2.5 dB will yield an increase in the output of only 1 dB, or for an input rise of 5 dB, the output will only rise by 2 dB, or as can be seen in the plot, an input change of 20 dB yields an output change of a little less than 10 dB (once the curve has reached its maximum slope). A compressor such as this would actually make

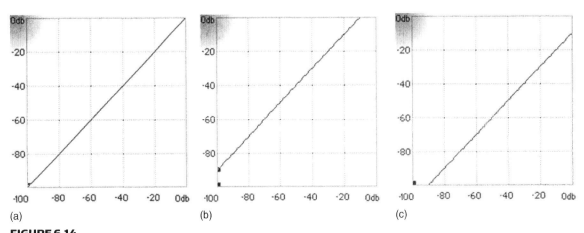

(a)   (b)   (c)

FIGURE 6.14
Three transfer curves: (a) unity-gain amplifier, (b) an amplifier with 10 dB gain and (c) amplifier with 10 dB loss (attenuation).

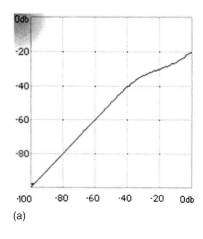
(a)

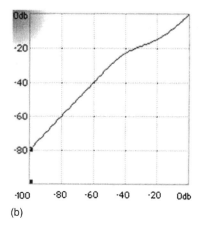
(b)

FIGURE 6.15
(a) Compressor with 2.5:1 ratio and −40 dBFS threshold (including a soft knee) and no gain makeup. (b) The same compressor with 20 dB gain makeup.

loud passages softer, because the output above the threshold is less than the input; this is always the case unless the compression is followed by gain makeup (a simple-gain amplifier after the compression section).

In Figure 6.15b, using makeup we can restore the gain so that a full level (0 dBFS) signal input will yield a full level signal output. For illustration purposes we show an amplifier with an extreme amount of gain, 20 dB, which would considerably amplify soft passages (below the threshold). In typical use, however, makeup gains are rarely more than 1–3 dB. Loud input passages from about −40 dBFS to about −15 dBFS are still amplified in this figure, but above about −15 dBFS, the curve slopes back to unity gain and resembles that of a linear amplifier. Far below the threshold, it's a fairly linear 20 dB amplifier and can have pretty low distortion because there is no gain reduction action. At full scale, 20 dB of gain makeup is summed with 20 dB of gain reduction yielding 0 dB total gain.

This particular compressor model's curve levels off toward a straight line above a certain amount of compression, so the ratio only holds true for the first 15–20 dB above the threshold. Other compressor models continue their steep slope, thus maintaining their ratio far above the threshold. There are as many varieties of compression shapes as there are brands of compressors and they all give different sounds. To get the greatest esthetic effect from any compressor, most of the music action must occur around the threshold point, where the curve's shape is changing; thus, a real-world compressor's threshold would likely be −20 to −10 dBFS or higher.

Knee

Figure 6.16 shows a very high ratio of 10:1; above the threshold the output is almost a horizontal line, which is a very severe compression, commonly called limiting.[17] The portion of the curve near the threshold is called the knee, which marks

---

[17]It is really a matter of degree, but most authorities call a compressor with a ratio of 10:1 or greater a limiter. The knee should also be very sharp for most effective limiting. Very few analog compressors have higher ratios, however, some digital limiters have been built with ratios of 1000:1 to prevent the minutest overload.

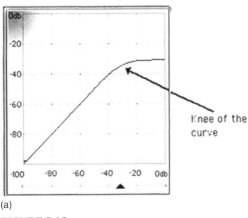

(a)

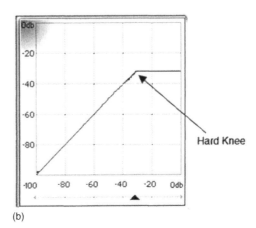

(b)

FIGURE 6.16
Compressor with (a) soft knee and (b) hard knee.

the transition between unity gain and compressed output. The term soft knee refers to a rounded knee shape or gentle transition (Figure 6.16a), and hard knee to a sharper shape (Figure 6.16b), where the compression reaches full ratio immediately above threshold. Soft knee can sweeten the sound of a compressor near the threshold. For those models of compressors that only have hard knees, lessening the ratio or raising the threshold, which will result in less action, can simulate some of the effect of a soft knee.

Attack and Release Times

Attack time is the time it takes for the compressor to implement full-gain reduction after the signal has crossed the threshold. Typical attack times used in music mastering range from 50 to 300 milliseconds (or longer on occasion), with the average used probably 100 milliseconds. Because digital compressors react with textbook speed, a digital compressor set to 100 milliseconds may sound similar to an analog compressor set to, say, 40 milliseconds; so it's probably better to remove all the labels on the knob (except slow and fast) and just listen! Release time, or recovery time, is how long it takes for the signal level to return to unity gain after it has dropped below the threshold. Typical release times used in music range 50–500 milliseconds or as much as a second or two, with the average around 150–250 milliseconds.[18]

The preview, or look-ahead function, allows very fast or even instantaneous (zero) attack time, which is especially useful in a peak limiter. In effect the unit has to react to the transient before it has occurred! This obviously requires a delay line, so there is no look-ahead in an analog processor. Every compressor has a sidechain, which is the control path (as opposed to the audio path), as illustrated in Figure 6.17.

---

[18]One manufacturer, DBX, measures release time in dB/second, which is probably more accurate, but I find it hard to get used to.

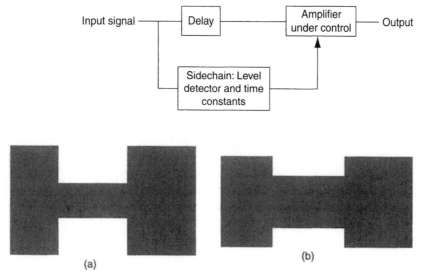

FIGURE 6.17
Look-ahead and
sidechain in a digital
compressor.

FIGURE 6.18
(a) A simple tone
burst from high to low
level and back. (b)
The same tone burst
passed through a
compressor with very
fast attack, high ratio
and fast release time.

The compressor places a time delay in the audio signal chain, but not in the sidechain, which allows the sidechain time to "anticipate" the leading edge of a transient and nip it in the bud. The delay time only has to be as long as the shortest transient we want to control plus the reaction time of the sidechain circuit. Look-ahead is really only relevant when desiring short attack times, since if we want a long attack then we probably also want to let initial transients pass through. Certainly analog compressors have gotten along splendidly without preview delay; in fact much of their sonic virtues come from their inability to stop initial transients. Because they contribute to the life and impact of a recording, I only remove sharp transients when they are audibly objectionable. The exceptions to this might be transients shorter than the ear can hear, which often occur in digital recording, and would prevent the program from having a higher intrinsic level. Removing these was the initial purpose of the brickwall limiter, designed to be short, quick and invisible.

Figure 6.18a shows the envelope shape of a simple tone burst, from a high level to a low one and back again.

Figure 6.18b shows the same tone burst passed through a compressor with a very fast attack, high ratio and fast release, and whose threshold is midway between the loud and soft signals. Note that the loud passages are instantly brought down, the soft passages are instantly brought up and there is less total dynamic range, as shown by the relative vertical heights (amplitudes).

Figure 6.19a shows the envelope of a compressor with a low ratio, slow attack time and a slow release time. Notice how the slow attack time of the compressor permits some of the original transient energy of the source to remain until the compressor kicks in, at which point the level is brought down. Then, when the signal drops below threshold, it takes a moment (the release time) during which the gain slowly

FIGURE 6.19
(a) A compressor with a slow ratio, slow attack time and slow release time. (b) The same tone burst passed through a compressor with very fast attack, high ratio and fast release time.

(a)

(b)

comes back up. A lot of the compression effect (the "sound" of the compressor) occurs during the critical release period, since except for the attack phase, the compressor has actually reduced gain of the high-level signal (Figure 6.19b).

Contrast this with the compressor at right, which has a much higher ratio, faster attack and very fast release time. The higher ratio clamps the high signal down farther, and with the fast release, as soon as the signal drops below threshold, the release time aggressively brings the level up. This type of fast action can make music sound squashed because it quickly brings down the loud and raises the soft passages.

The essential fact here is that (downward) compressors take the loudest passages down. Gain makeup allows the average level to be raised but the loudest passages end up proportionally lower. In mixing, this allows the engineer to mix an instrument either at a lower level without losing its low passages or at a louder level without having intermittent loud passages interfere with hearing other instruments. Although a compressor can add or enhance punch in mastering, since its essential mechanism is to reduce the partial loudness of peaks, if overused or if too many transient attacks are softened this can actually produce the opposite effect.[19] So a standard compressor should ideally be applied when loud passages in a recording sound problematic and/or when some parts sound too loud or too soft. Remember that manually raising the soft passages (avoiding the processor entirely) can leave the loud passages alone and yield a more impacting production. In my book, *Mastering Audio: The Art and the Science*, we investigate how upward compression can increase soft- and mid-level passages with little effect on the important loud ones, a technique which can produce a recording that is dynamic, loud and has impact at low levels.

Here is another variation, a compressor with a release delay.

A release delay control allows more flexibility in painting the sound character. Very few compressors provide this

FIGURE 6.20
Output of a compressor with a low ratio, slow attack time, slow release time plus release delay.

[19]Punch comes directly from microdynamic contrast, especially at bass frequencies. The level has to dip just a little in order for it to come back up and "hit you"; without a counterswing there can be no swing. In general, punch is achieved during the recording and mixing process, and frequently it is lost during mastering if the engineer is not careful.

©2009 Bob Katz

facility but it is useful when we want to retain more of the natural sound of the instrument(s), and not exaggerate its sustain when the signal instantly goes soft, or reduce "breathing" or hissing effects when the source is noisy.

Figure 6.21 illustrates what happens when the attack and release times are much too fast.

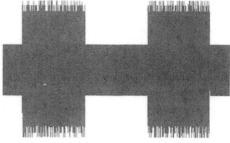

FIGURE 6.21
When the combination of attack and release times are extremely fast (typically <50 milliseconds), a compressor can produce severe distortion, as it tries to follow the individual frequencies (waves) instead of the general envelope shape of the music.

The distortion shown here is caused by the compressor's action being so fast that it follows the shape of the low-frequency waveform rather than the overall envelope of the music. This problem can occur with release times shorter than about 50 milliseconds and correspondingly short attack times.

MICRODYNAMIC MANIPULATION: ADJUSTING THE IMPACT OF MUSIC WITH A (DOWNWARD) COMPRESSOR

The Engineer As Artist

Compressors, expanders and limiters form the foundation of modern-day recording, mixing and mastering. With the right device engineers can make a recording sound more or less percussive, more or less punchy, more or less bouncy, or simply good or bad, mediocre or excellent.

Used in skilled hands, compression has produced some wonderful recordings. A skilled engineer may intentionally use creative compression to paint a sound and form new special effects; and a lot of contemporary music genres are based on the sound of compression, both in mixing and in mastering, from Disco to Rap to Heavy Metal. The key words here though are intent and skill. Surprisingly, however, some engineer/artists don't know what uncompressed, natural-sounding audio sounds like. Although more and more music is created in the studio control room, I think it's good to learn how to capture natural sound before moving into the abstract. Picasso was a creative genius, but he approached his art systematically, first mastering the natural plastic arts before moving into his cubist period. Similarly, it's a good practice to know the real sound of instruments. Recording a well-balanced group in a good acoustic space with just two mikes is not only a lot of work but also a lot of fun! Before multitracking was invented, there was much less need for compression. But because close miking exaggerates the natural dynamics of instruments and vocals, compressors were increasingly employed to control those instruments whose dynamics were severely altered by this microphone technique (e.g., vocals and acoustic bass). When modern music began to emphasize rhythm, many instruments began to get masked by the rhythmic energy, inspiring the creative possibilities of compressors and a totally new style of recording and mixing. The advent of the SSL console, with a compressor on every individual channel, changed the sound of recorded music forever.

COMPRESSION AND LIMITING IN MASTERING

Mastering requires us to develop new skills as it is concerned with overall mixes rather than individual instruments. Most mastering engineers use compressors to intentionally change sound. Compression is a tool that can change the inner dynamics of music (e.g., add punch to low- and mid-level passages, enhance rhythmic movement or make a stronger musical message). But limiters generally change sound very little, simply enable it to be louder.[20] That's why they are used more often in mastering than in mixing. Even the best limiter is not perfectly inaudible, softening the transients and even fattening the sound slightly.

BBC research in the 1940s demonstrated that distortion shorter than about 6–10 milliseconds is fairly inaudible, hence the 6 milliseconds integration time of the BBC PPM meter. But this result reflected also the then-current state of technology: with digital recording and solid-state equipment, some transient distortion as short as 1 millisecond will audibly change the sound of the initial transient, particularly for instruments such as piano. With good equipment and mastering technique, program material with a peak-to-average ratio of 18–20 dB can often be reduced to about 14 dB with little effect on the clarity of the sound but some reduction in transient impact. That's one of the reasons 30 IPS analog tape is so desirable: it has this limiting function built in. A rule of thumb is that short-duration (a few milliseconds) transients of unprocessed digital sources can often be transparently reduced by 2 dB, in rare cases as much as 6 dB with little effect on the sound; however, this cannot be done with analog tape sources, which have already lost the short-duration transients.[21] Any further transient reduction by compression or limiting, whatever its desirability, will not be transparent. Limiting distortion is especially audible on material that already has little peak information because a limiter is not designed to work on the RMS portion of the music and they can sound harsh when pushed into the RMS region. So the less limiting, the more snappy the sound.

In an ideal mastering session, if a limiter is used at all, it should only be acting on occasional inaudible peaks, or perhaps a bit more if we like the slight fattening effect. A manual for a certain digital limiter reads "For best results, start out with a threshold of −6 dBFS." This is like saying "always put a teaspoon of salt and pepper on your food before tasting it." One modern R&B album is so overlimited that the bass drum punches a hole in the vocal on every attack; I doubt this is artistically desirable, although in some genres of music, this technique is used for artistic means.

It is a common misconception that a limiter is a peak protection device for mastering. It may be used as such in radio broadcasting (to protect the transmitter) or

---

[20]As with compressors, it is the gain makeup process that permits the output of a limiter to be louder. When the peaks have been brought down, there is room to bring the average level up without overloading. For example, snare drum hits that stick out above the average can be softened by the peak limiter (if this change in sound proves desirable) and the average level can then be raised.

[21]Limiter release time is important: the faster the release time, the greater the distortion, which is why the only successful limiters which use extra fast release times have auto-release control, which slows down the release time if the duration of the limiting is greater than a few milliseconds. The effective release time of an auto-release circuit can be as short as a couple of milliseconds and as long as 50–150 milliseconds. If limiting a very short (invisible) transient, the release time can be made very short.

sound reinforcement (to protect loudspeakers), but in mastering (or mixing), the engineer has total control over his/her levels and makes the choice whether to turn them down or raise them and use a peak limiter.

The World's Most Transparent Digital Limiter

The most transparent limiter is to use no limiter at all! If there is a very short peak (transient) overload, for example, a drumbeat within a section which needs to be made louder, a skilled mastering engineer can use digital audio workstation's (DAW's) editor to perform a short-duration gain drop that can be quite inaudible. This manual limiting technique allows us to raise a song's apparent loudness without inducing distortion from a digital limiter, so it is the first process to consider when working with open-sounding music that can be ruined by too much processing. We can often get away with 1–3 dB manual limiting typically for a duration of less than 3 milliseconds. But, longer-duration manual gain drops will affect the sound as much as or more than a good digital limiter.

Equal-Loudness Comparisons

As loudness has such an effect on judgment, it is very important to make comparisons at equal apparent loudness. If played louder, during an instant A/B comparison the processed version may initially seem to sound better but long-term listeners prefer a less fatiguing sound that "breathes." When we compare them at a matched loudness, we may be surprised to discover that the processing is making the sound worse, and the "improvement" was an illusion. When making an album at "competitive loudness level," I'm relieved if I find that the mastering has not degraded the sound of the program and ecstatic if it has improved it.

The Nitty-Gritty: Compression in Music Mastering

Consider this rhythmic passage representing a piece of modern pop music:

shooby dooby doo WOP...

shooby dooby doo WOP...

shooby dooby doo WOP

The accent point in this rhythm comes on the backbeat (WOP), often a snare drum hit. If we strongly compress this music piece, it might change to:

SHOOBY DOOBY DOO WOP...

SHOOBY DOOBY DOO WOP...

SHOOBY DOOBY DOO WOP

This completely removes the accented feel from the music, which is probably counterproductive.

A light amount of compression might accomplish this …

shooby dooby doo WOP...

shooby dooby doo WOP...

shooby dooby doo WOP

which could be just what the doctor ordered for this music as strengthening the subaccents may give the music more interest. Unless we're trying for a special effect, or purposely creating an unnatural sound, it's counterproductive to go against the natural dynamics of music. (Like the TV weatherperson who puts an accent on the wrong syllable because they've been taught to "punch" every sentence: "The weather FOR tomorrow will be cloudy.") Much of hip-hop music, for example, is intentionally unnatural—anything goes, including the eradication of any resemblance to the attacks and decays of real musical instruments.

"Musical" compression requires careful adjustment of compressor parameters. If the attack time is too short, the snare drum's initial transient could be softened, losing the main accent and defeating the whole purpose of the compression. If the release time is too long, then the compressor won't recover fast enough from the gain reduction of the main accent to bring up the subaccent (listen and watch the bounce of the gain reduction meter). If the release time is too fast, the sound will begin to distort. If the combination of attack and release times is not ideal for the rhythm of the music, the sound will be "wimpy loud" instead of "punchy loud." Getting it right is a delicate process, requiring time, experience, skill and an excellent monitoring system.

One way engineers start to compress in order to help obtain punch or attitude is first to find the threshold. Using a very high ratio (say 4:1) and very fast release time (say 100 milliseconds) the engineer will then adjust the threshold until the gain reduction meter bounces as the offending "syllables" pass by and the bounce can be heard. This ensures that the threshold is optimally placed around the musical accents the engineer wants to manipulate, the "action point" of the music. Then the ratio will be reduced to a very low setting (say 1.2:1) and the release time raised to about 250 milliseconds to start. From then on, it's a matter of fine tuning attack, release and ratio, with possibly a readjustment of the threshold. The object is to put the threshold in between the lower and higher dynamics, creating a constant alternation between high and low (or no) compression within the music.

Don't get fooled by your eyes, as many compressors' meters are too slow, they go into gain reduction before the meters move, so only 1 dB of metered-gain reduction can mean a lot.

Typical Ratios and Thresholds

When working on microdynamics in the above fashion, compression ratios most commonly used in mastering are from about 1.5:1 to about 3:1, and typical thresholds in the −20 to −10 dBFS range. But there is no rule; some engineers get great results with ratios of 5:1, whereas a delicate painting might require a ratio as small as 1.01:1 or a threshold of −3 dBFS. One trick they use to compress as inaudibly as possible is to employ an extremely light ratio, say 1.01 to 1.1 and a very low threshold, perhaps as low as −30 or −40 dBFS, starting well below where the action is. In this case, the compressor is not bouncing on the syllables but rather giving a gentle, continuous form of macrodynamic reduction. We may choose a low ratio to lightly control a recording that's too jumpy or to give a recording some much needed body. It's unusual to see such low ratios in tracking and mixing but very common in mastering, partly because with full program material, larger ratios may create breathing, pumping or other artifacts.

Compressors with Unique Characteristics

We have noted before that every brand of processor (both compressors and expanders) has its own unique characteristics and sound. Part of the fun of mastering (and mixing) is discovering the specialities of different compressors. Even with the same settings, some are smooth, some punchy, some nicely fatten the sound and others make it brighter, harder or more percussive. This is often due to differences in the curve or acceleration of the time constants, how the device recovers from gain reduction, whether the gain returns to unity on a linear, logarithmic or even an irregular curve. Design engineers spend much research time discovering the bases of a compressor's sound.

Analog compressor designers choose from several styles of gain manipulation. The most common are optical (abbreviated opto), VCA (voltage controlled amplifier), Vari-Mu, PWM (pulse width modulation) and their various subcategories. Digital designers may emulate their characteristics, as in the Waves Renaissance series of digital compressors which have both opto and electro modes. In opto position, the release time slows down for the last portion of the release, whereas in electro it accelerates. Electro can yield a more aggressive sound, whereas opto is good for gentle, easy-going purposes. Analog optical compressors are great on vocals in tracking or mixing, not as good for mastering of overall program material because they are generally too slow. However, digital opto models can be faster than their analog counterparts. Or add additional emulation features, like supplementary low-frequency harmonics to change tonality, as in Waves "warm" setting of the Renaissance Compressor.[22] Note that the addition of harmonics slightly compresses sound by reducing the peak-to-average ratio.

While generally analog optical models are more suitable for "gentle" mastering, one model, the Pendulum OCL-2, has a proprietary optical sensor whose reaction time is much faster than others on the market as well as a very transparent tube circuit that can provide very subtle warming. This makes the OCL-2 perhaps the only optical analog compressor with the gentleness of optical (useful for adding body) but also capable of providing a bit of punch. However, it is not as fast as a VCA- or PWM-based compressor, which may be necessary to help achieve the right results in rock and roll. In my opinion, closest to a jack-of-all-trades is the Cranesong Trakker, a compressor that can emulate the tonality and speed characteristics of several different types of compressors and though solid state, some of the warming characteristics associated with tubes. The only downside of the Trakker is the learning curve (the best way to learn is to experiment with each of the presets). Another style of compressor is the Manley Vari-Mu whose ratio varies with dynamics but its action may be a bit slow for music with fairly fast dynamics.

---

[22]Waves named the alternative setting (when the extra harmonics are turned off), "smooth," because they did not want to prejudice the user with a term such as "cool." But this has nothing to do with "smooth" so in retrospect they should simply have labeled one position of the switch "warm" with no label in the other position.

Sidechain Manipulation

Most of the time the sidechain (control path) is identical to the audio signal but interesting things can happen when it is not. For example, in a stereo or multichannel compressor, each channel has its own sidechain, but it is possible to feed or link all sidechains from one channel's signal. By linking the sidechains, one channel controls the gain reduction of both equally. The linking switch prevents image wandering, for if a drum hits much louder in one channel than the other, with an unlinked sidechain, the image will momentarily move toward the opposite channel. When unlinked the box operates as two independent mono compressors. In some models, unlinked is labeled dual and linked is labeled stereo. In multichannel compressors, there may be a separate sidechain for front and surround, or all channels may be linkable.

Often sidechains are fed an equalized signal. Perhaps the most popular sidechain EQ is a high pass-filtered signal, which helps prevent the bass drum from pushing down (or modulating) the rest of the music.

With a sidechain boost in the upper mid or low treble range, the compressor becomes a de-esser or it can be tuned to deal with troublesome cymbals. The only problem with sidechain-based de-essing is the entire range of audio frequencies is brought down whenever an "s" goes by, so generally the gain reduction has to be kept subtle, no more than a decibel or two, unless you are looking for a special effect.

Another use of sidechains is for ducking, where a completely separate signal is fed to the sidechain than the audio path. This is used mostly for special effects or in speech radio—when an announcer speaks, the music underneath can be automatically reduced. If the release time is set slow, then the music will slowly return to normal gain when he/she stops speaking.

Multiband Processing: Advantages and Disadvantages

Splitting a compressor's signal into multiple bands (and multiple sidechains) avoids the problem of modulation with a single sidechain as compression action in one band will not affect another band. For example, the vocal will not pull down the bass drum (or vice versa). This is perhaps the biggest selling point of multiband because with the same amount of gain reduction it can sound superior to wideband or sidechain equalization. Or a higher amount of compression and average level can be achieved in a multiband with fewer interaction artifacts.

Another advantage is that high-frequency transients can be left unaffected while compressing the mid-range more severely, producing a brighter, snappier sound. However, loud action in one frequency band can dynamically change the tonality, producing an uncohesive sound especially if all the bands are moving in different amounts throughout the song but even this property of multiband can be put to advantage. When slightly more compression is applied in the high-frequency region, the sound gets duller as it gets louder, which is a way to construct an analog tape simulator, to sweeten digitally recorded material, or to soften distortion that gets harsh when it gets loud.

Multiband units make good de-essers. Sibilance can be controlled by using selective compression in the 3–9 kHz range (the actual frequency has to be tuned by listening to the vocalist). Try a very fast attack, medium release, crest factor set to peak (to be explained) and a narrow bandwidth for the active band. The Weiss DS1-Mk2 is hands down the best-sounding mastering de-esser I've encountered probably due to its peak sensitive compression, linear-phase band-split filters, dual-speed release and very effective presets.

The multiband device's virtues permit louder average levels than were previously achievable—making it, in my opinion, the most powerful but also potentially the most deadly audio process that's ever been invented. I say "deadly" because multiband compression helps fuel the loudness race (or loudness wars as previously mentioned). The technique has been hyped as a cure for all ills (which it is not) and it can easily produce very unmusical sound or take a mix where it doesn't need to be. But it can also be used to help improve ("repair") a bad mix when a re-mix is not possible, and some mastering engineers become experts at this technique. For example, I once received a rap project that was somehow mixed with very low vocal, extremely loud percussion and bass drum. A re-mix was not possible but by compressing and then raising the level of the frequencies in the vocal range (circa 250 Hz) I was able to rebalance the piece and turn the vocal up. Just don't fool yourself into believing that toning down loud instruments through multiband compression is the same as a re-mix.

Multiband processing was probably first introduced by TC Electronic in their M5000, then in their ubiquitous Finalizer and brought to great sophistication and versatility in their System 6000 with the MD4, perhaps the first multiband compressor to introduce linear-phase band-split filters.[23] But for most downward compression purposes multiple bands are rarely needed; one or two bands are usually enough. The Weiss DS1-Mk2 has one active band; the compressed signal can be isolated to one frequency range and the rest of the spectrum left unaffected. Most of the compression I perform in mastering is with a full-band compressor, or a full-band compressor with a high-pass sidechain or the Weiss with the active band above some low frequency. Regardless of this advantage, rarely do even hip-hop recordings need more than two bands to sound punchy and strong. I use a downward compressor with more than two active bands in my mastering a small percentage of the time, when multiple bands have been a lifesaver on bad mixes, but let us not forget that the key to a great master is to start with a great mix![24]

Before trying multiband, first:

- See if simply raising the attack time in a one-band compressor permits sufficient transient energy to come through.
- Try using few bands, only two if possible. This reduces potential "phasey" artifacts if the filters in the compressor are not linear phase.

---

[23]The Weiss DS1-Mk2 uses a linear-phase band-split filter, but it only has one active band, so it cannot be properly called a multiband compressor.
[24]A variation of this quote is in Owsinsky, Bobby. *Mastering Engineer's Handbook.*

Equalization or Multiband Compression?

When multiband processing is used, the line between equalization and dynamics processing becomes nebulous, because the output levels of each band form a basic equalizer. Standard equalization should be applied when instruments at all levels need alteration, or a multiband compression can be employed to provide spectral balancing at different levels. This is a form of dynamic equalization, and so, depending on the point of view of the engineer, a multiband compressor can be looked upon as a dynamic equalizer.

If I'm already using a multiband unit, I make my first pass at equalization with the outputs (makeup gains) of each band. Multiband compression and equalization work hand-in-hand. Tonal balance will be affected by the crossover frequencies, the amount of compression and the makeup gain of each band. As we know, the more compression, the duller the sound, so I first try to solve this problem by using less compression, or altering the attack time of the high-frequency compressor, and as a last resort, I use the high-frequency band's makeup gain or an equalizer to restore the high-frequency balance.

Emulation vs. Convolution

Digital compressor designers have the choice of emulating the transfer characteristics of a source compressor and implementing them in DSP, or sampling the source compressor and convolving that sampled characteristics with the incoming audio. Convolution works well with reverbs and equalizers, but in my opinion, I have yet not heard a successful convolution-based digital compressor.

Fancy Compressor Controls

Some compressors provide a crest factor control, usually expressed in decibels, or a range from RMS (or full average) to quasi-peak through to full peak. What this means is that the compressor can be set to act on either the average parts of the music, the peak parts, or somewhere in between. Ostensibly, compressors with RMS characteristics sound more natural as they correspond with the ear's sense of loudness, but one of the best-sounding compressors I know is peak sensing. When the crest factor control is set to peak, short transients tend to control the action, and at RMS, more continuous sounds control it. The TC MD4 has a continuously variable crest factor control. For most music work I tend to leave it at RMS but for de-essing and for better control of transients (such as a too-loud snare drum) I move it closer to peak.

The Weiss model DS1-Mk2 is the first dynamics processor I've encountered with two different release time constants, release fast and release slow. The user sets a threshold of average transient duration, such as 80 milliseconds, above which a sound movement is called slow and below which it is called fast. Instantaneous transients can be given a faster release time, but sustained sounds a slower one, which results in a more natural-sounding compression, especially with heavy compression. Indicator lights on the front panel make adjusting this to a snap.

The Roger Nichols Dynamizer is a dynamics processor with multiple thresholds. Compressors with multi-thresholds can be simulated by running two (or more) compressors in series. For example, the first compressor performs a gentle overall compression with a low threshold and ratio, and the second more aggressively controls some offensive peaks at high levels. It's not uncommon to run several dynamics processors in series in mastering, each doing a small part of the job. To keep a peak limiter from clamping down on the signal too much, I may precede it with some form of preconditioner, which may be an analog compressor that can gently soften a transient so that the limiter which follows doesn't have to work as hard.

Clipping, Soft Clipping, and Oversampled Clipping

Clipping is the result of attempting to raise the level higher than 0 dBFS, producing a square wave, a severe form of distortion. Clippers are devices which electronically cut momentary peaks out of the waveform to allow the overall level to be raised. Soft clipping attempts to do this with less distortion. I don't like the quality of distortion produced by clipping or soft clipping, at least at a 44.1 kHz sample rate and I believe there are better approaches. The first is not to raise the level at all, for many CDs are already too hot for their own good; the second is to use a brickwall limiter, as subtly as possible.

Compression, Stereo Image and Depth

Compressors tend to amplify the mono information in a recording, which affects stereo width. Compression also deteriorates the depth in a recording as it brings up the inner voices in musical material. Instruments that were in the back of the ensemble are brought forward, and the ambience, depth, width and space are degraded. Not every instrument should be "up front." Pay attention to these effects when you compare processed vs. unprocessed and listen for a long enough time to absorb the subtle differences. Variety is the spice of life. As always, make sure the cure isn't worse than the disease.

The Mastering Engineer's Dilemma

Without compressors in CD changers and in cars, it is extremely difficult for the mastering engineer to fulfill the needs of both casual and critical listeners. It is our duty to satisfy the producer and the needs of the listeners, so we should continue to use the amount of compression necessary to make a recording sound good at home. But try to avoid using more compression than is required for home listening. If compromises have to be made for car or casual play, try transparent-sounding techniques such as parallel compression, which satisfy even critical listeners and offer their own unique advantages.

Parallel compression is a process where the signal is split into two, one signal often remains clean (or can be gently compressed) whilst the other signal is compressed pretty hard. The clean and the compressed signal can be then "blended" until the right effect is achieved. An equivalent would be a wet/dry control on the compressor, such as is now sported by Flux's Solera compressor, a version of which is often employed by Pyramix DAW users.

FURTHER READING

This edited Chapter is taken from Katz, B. (2007). *Mastering Audio: The Art and the Science*. Focal Press, Oxford

Nisbett, A. (2003). *The Sound Studio: Audio Techniques for Radio, Television Film and Recording*. Focal Press, Oxford.

Chapter 7
Putting the Album Together

Bob Katz

In This Chapter

In this chapter, Bob Katz speaks about the way in which a collection of seemingly disparate songs can be placed together by the mastering engineer such as himself to make a powerful statement as an album.

INTRODUCTION

Although we have now entered an era of digital downloads, with an increased emphasis on singles combined with a perceived shorter attention span of the listening public, the album is still an important music medium. Sergeant Pepper is often cited as the first rock and roll concept album (i.e., an elaborately designed album organized around a central theme that makes the music more than a simple collection of songs). This started a trend that many assume has more or less died. But is the concept album really dead? I'm not so sure; I treat every album that comes for mastering as a concept album, even if it doesn't have a fancy theme, artwork or gatefold. The way the songs are spaced and leveled contributes greatly to the listener's emotional response and overall enjoyment. It is possible to turn a good album into a great album just by choosing the right song order and of course the converse is also true.

HOW TO PUT AN ALBUM IN ORDER

Sequencing is something of an art. (Sequencing here is not to be confused with the arrangement of MIDI parts in a Digital Audio Workstations, also referred to as DAWs, such as Logic or Cubase.) Sometimes, the musicians making an album have a good idea of the song order they'd like to use but many people need help with this task. Traditionally, the label's A&R person would help put the album in order, but with independent productions, that service is not always available and so it falls to the producer, or someone experienced, politically "neutral"[1] and so an experienced mastering engineer is well placed to provide useful guidance during this process.

My advice is to avoid over-intellectualizing. One musician decided to order his album by the themes of the lyrics; he started with all the songs about love, followed by the songs about hate, and finally the songs about reconciliation. It was a musical disaster. The beginning of his album sounded musically repetitive, because all his love songs tended to use the same style, and furthermore, the progression of intellectual ideas simply was not obvious to the average listener, who primarily reacted to the musical changes. Even when the listener got the underlying point, it didn't contribute much to the enjoyment of the album. Listening to music is first and foremost an emotional experience. If we were dealing with lyrics without music (poetry), perhaps the intellectual order would be best, but the intellectual point of the album will still come through even if the songs are organized for primarily musical reasons.

Before considering the order of the album, it's important to have its gestalt in mind: its sound, its feel, its ups and downs. I like to think of an album in terms of a concert. Concerts are usually organized into sets, with pauses between the sets when the artist can catch her breath, talk briefly to the audience, and prepare the audience for the mood of the next set. On an album, a set typically consists of three or four songs but can be as short as one. Usually the space between sets is a little greater than the typical space between the songs of a set, to establish a breather or mood change. Sometimes there can be a long segue (crossfade) between the last

---

[1]A neutral producer helps prevent the more me syndrome, commonly encountered with self-produced albums. He/she also helps achieve the goal of a cohesive album rather than a collection of different band member's tunes.

song of a set and the first of the next. This basic principle applies to all kinds of music, vocal and instrumentals; it is analogous to the spacing in a classical music album, shorter ones between movements of a single composition and longer ones between the compositions themselves.

To make the job of organizing the sets easier, the mastering engineer or the artist can prepare a rough CD of all the songs, or a playlist on iTunes® or better still a DAW, to allow instant playback of all the candidates. This is a lot easier than it was in the days of analog tape! Then make a simple list, describing each song's characteristics in one or two words or symbols, such as up-tempo, mid-tempo, ballad. Sometimes I'll give letter grades to indicate which songs are the best performed, most exciting or interesting, trying to place some of the highest grade songs early in the order for a good first impression. I may note the key of the song, although this is usually secondary compared to its mood and how it kicks off. If there's a bothersome clash in keys, sometimes more spacing helps to clear the ear, or else I exchange that song with one that has a similar feel and compatible key.

The opening track is the most important; it sets the tone for the whole album and must favorably prejudice the listener. It doesn't have to be the hit or the single but most frequently is up-tempo and establishes the excitement of the album. Even if it's an album of ballads, the first song should be the one that is most likely to engage the listener's emotions.

If the first song was exciting, we usually try to extend the mood, keep things moving like a concert, by a short space, with an up- or mid-tempo follow-up. Then, it's a matter of deciding when to take the audience down for a breather. Shall it be a three- or four-song set? I examine the other available songs, then decide if it will be a progression of a mid-tempo or fast third song followed by a relaxed fourth, or end with a nice relaxed third song.

At this point, I have provisional track numbers penciled next to the candidates for the first set. I play the beginning of the first song to see how it works as an opener, then skip to the last 30 or 40 seconds, play it out and jump to the start of the second song. If this musical transition doesn't work, then the sequence is faulty regardless of how compatible the two songs seem to be. That's why transitions can let us join different musical feels; an up-tempo song that winds down gently at the end can easily lead to a ballad. If the set doesn't flow, I substitute songs until it does.

I then check off the songs already used and pick candidates for the second set, usually starting with another up-tempo in a similar "concert" pattern. This can be reversed; some sets may begin with a ballad and end with a rip-roaring number, largely depending on the ending mood from the previous set. A set can also be a roller-coaster ride, depending on the mood we want to create, but when you consider the album in terms of sets, it all becomes a lot easier to organize. By the way, the ultimate listener doesn't usually realize there are sets; our work should be subliminal.[2] As the list gets filled up, it becomes a jigsaw puzzle to make the remaining

---

[2]Though everyone has their favorite album transition like in Sergeant Pepper between the rooster crow and Good Morning. The producer claims it was a fortuitous accident.

pieces fit. Perhaps the third or fourth set doesn't work quite as well as the first. Perhaps one of the transitions is clashing, even if we increase the spacing. At that point, I may try a one-song set, or try to place this problem song into an earlier set, either replacing a song or adding to the earlier set.

The Odd Man Out

One song may just not fit well musically with the rest. For a Brazilian samba album, the artist also recorded a semi-rock blues number. She said everyone loved this song in Brazil, so we couldn't excise it from the album but stylistically it did not gel as a part of any set. At first I suggested putting it last as a "bonus track," but this ruined the original album ending, which was a beautiful, introspective song that really did belong at the end. Eventually, we found a place for the offender near the middle of the sequence, as a one-song set, with a long enough pause before and after. It served as a bridge between the two halves of the album.

The Right Kind of Ending

So, how to end the album? What is the final encore in a concert? It's almost never a big, up-tempo number, because the audience always cries "more, more, more." You've got to leave them in a relaxed, comfortable "goodbye mood," otherwise you'll be playing encores forever. That's why the last encore is usually an intimate number, or a solo, with fewer members of the band. The same principle applies with an album. I usually try to create a climax, followed by a dénouement. The climax is obviously an exciting song that ends with a nice peak. This followed by one or two easy-going songs to close out the album. When I find the perfect sequence, it's a real treat!

SPACING THE ALBUM

The next thing to remember is never to count the seconds between songs. Experienced producers know that the old "4-second," "3-second," or "2-second" rule really does not apply, although it is clear that album track spacing has gotten shorter over the past 50 years, along with the increased pace of daily life. The correct space between songs can never be accurately measured, for different people start counting at different times depending on when they think the decay is over. Counting from the beginning of true silence, the computer may objectively say that a space is only 1 second, but the ear may think it's closer to 2.5. So don't count—just listen. As a general rule, the space between two fast songs is usually short, between a fast and a slow song it is medium length, and between a slow and a fast song it is usually long. After a fadeout, the space is usually very short, because the listener in a noisy room or car doesn't notice the tail of a fade-out. Often we have to shorten fade-outs and make segues[3] or the space will seem overextended, especially in the

---

[3]Segue (pronounced seg-way)—a crossfade or overlap of two elements. Webster's—proceed without interruption. Italian—seguire, to follow.

car. Perception of appropriate spacing can depend on the mood of the producer and the time of day. If a mastering engineer spaces an album in the morning when she's relaxed, it almost always sounds more leisurely than one which has been paced in the afternoon, when her heart generally beats faster. To avoid being unduly influenced by these external factors engineers should try not to make spaces too short when they're in an energetic mood or too long a space when they're very relaxed.

The overall pace of an album is also affected by intertrack spacing. We probably want the first set to be exciting and so control the pace using shorter spaces within the first set and then slightly longer spaces thereafter. An interesting observation is that our sense of timing is relative, if we begin with very tight spaces and then revert to "normal" spaces, these seem too long. Manipulate spaces to produce special effects—surprises, super-quick and super-long pauses make great effects. One client wanted to have a long space in the middle of his CD, about 8–10 seconds, to simulate the change of sides of an LP. Respecting the input of a creative individual I tried the super-long space and it worked! This was largely due to his choices of songs and the order. The set that began side two had a significantly different feel, and the long space helped to set it off, like a concert intermission.

Some people think that it's sufficient to play the last 30 seconds of a tune in order to judge the space before the next. But if you play an entire exciting song, you will most assuredly need more space to catch your breath before the next one can start. To avoid playing a whole song, we try to anticipate this effect by using a slightly longer space; when we play the album through we'll know if we were successful. One technique for judging a space is to cut it shorter and shorter until it is obviously too short and then add just the scintilla necessary to make it sound "just right," especially knowing that it will seem longer on a domestic hi-fi. Another type of space is to make the downbeat of the next song be in time with the rhythm of the previous. This can sound very nice but not if overused.

We didn't have this luxury in the days of analog tape, and it's interesting to note that when an LP master comes in for conversion to CD the spaces always seem too long. One reason, as I've said before, is the current quicker pace of life but the other is that vinyl and tape noise acts as a filler. When there's dead silence between tracks, spaces always seem longer. I may remove 2 or more seconds out of an LP space and it will be just fine on CD.

PQ CODING
PQ Offsets

Most authorities recommend placing a track start mark (called Index 1) at least 5 SMPTE frames before the downbeat to accommodate slow-cuing CD players. This is approximately 12 CD frames, 160 milliseconds (each CD frame is 1/75 second). Modern DAW systems typically used by mastering engineers can automatically apply these offsets and show the PQ codes as they will appear on the disc. Sophisticated DAWs such as these let you rehearse the effect of cuing with or

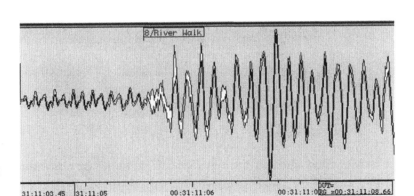

FIGURE 7.1
Track mark placed very tight to the downbeat with no offset to avoid hearing talking which comes before the mark.

without the offsets, critical when the cue has to be very tight.[4] For example, when the previous song is crossfading into the next, if we do not place the track mark extremely close to the downbeat of the next song the CD player may play a piece of the previous sound. I may accept as little as 2 (occasionally 1) SMPTE frames, which risks that a slow-cuing player will miss the downbeat. Figure 7.1 is an example of a live album with the track mark located nearly on top of the downbeat to avoid the spoken introduction. Some players clip the downbeat but on this CD it was less of a problem than hearing the previous sound.

Spaces and PQ (Track) Coding

Index 0 is an optional mark between the tracks which defines the end of the previous track; the CD player's time display begins to count backward up to the Index 1. This is called the pause time, a misleading term, for there is no requirement for silence and in fact, Index 0 can be 0 seconds. I recommend that mastering engineers should normalize Index Zeros shorter than 2 to 0 seconds to keep the player's time display from glitching. This doesn't mean that musical spaces cannot be as short as you want them to be, it only means there will be no official pause between tracks. When Index 0 is 0 seconds, the player interprets Index 1 as the end mark of the previous track and the start of the next.

Hiding Information in the Gap

When a cut from a concert album is played on the radio, it's often desirable to cue the tune on the downbeat, but the listener at home wants to hear the atmosphere between cuts and maybe the artists' introductions. To accomplish this dual feat, the creative mastering engineer can place Index 0 and Index 1 times as in Figure 7.2.

---

[4]Some DAWs print PQ lists with both the offset and non-offset PQ times, but the plant only needs to see the offset times, which appear on the CD master. The other standard offsets used in mastering programs are to place the first track mark 2 seconds before the beginning of the music, a Redbook recommendation; an end mark a few frames after each song end, to keep some players from going into premature muting; and to place the last end mark 2 seconds after the end of music, so the player can stop spinning without losing the last sound.

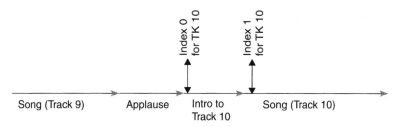

FIGURE 7.2
Within the Red Book standard are the Index markers that can be very effective when dealing with audience applause in live albums.

In this example, the song for track 9 ends with applause, and the official end of song 9 is at the Index 0. There is sound in the pause or gap time between Index 0 and Index 1; this permits consumer choice or the CD player's random play function to ignore the boring or irrelevant parts. Similarly, the introductions, count offs, sticks and so on for songs on any kind of album can be placed in the gap so they will not be heard on the radio or in random play. Note that the pause time does not count as part of the official length of either track (keeping royalty costs down!). Unfortunately, this functionality of the CD standard is eroding, hindering artistry that we have enjoyed for over 20 years. iTunes® and some primitive CD players do not read Index 0; they treat the introduction or the countoffs as the end of the previous track, producing some incongruous results in random play. Furthermore, The iPod® (until the 5th generation and iTunes® version 7) and some other players briefly mute at each track start, breaking up the continuity of a continuous album. After version 7, Apple's music players can play gapless albums without glitches but only if the user explicitly marks the tune as part of a gapless album.[5] Regardless, I always PQ code masters assuming they will be played on CD players that respect the standard; there is little other choice.

In the same vein, it pays to be vigilant because some CDR duplicators will mute the pause audio, sometimes even taking many seconds OUT and putting just 2 blank seconds IN. These copiers were found to be copying in Track At Once Mode, rather than Disc At -at- Once, instead of simply cloning the disc.[6] Imagine the classic Pink Floyd *The Wall*, which has continuous sound, being gapped by accident at the plant.

Red Book Limits

The Sony/Philips Red Book specifies all the parameters for an audio CD. A CD may have up to 99 tracks, whose minimum length is 4 seconds, and each of these tracks may have up to 99 indexes (or subindices). Rarely do we code CDs with indexes since many players do not now support them and most people don't know how to use them. Classical engineers who used to code movements with indexes are now using a track mark for each succeeding piece. There is no standard CD length; maximum length can be stretched to 80 minutes if the plant tightens the disc spirals to the minimum Red Book tolerance—but not all players can play the outer tracks of

---

[5]This is a fundamental limitation of coded formats like MP3 that cannot be easily edited or crossfaded. iTunes® must have engineered a workaround to enable gapless albums.
[6]Thanks to Dan Stout for this information as viewed on the Mastering Webboard.

these discs. Individual plants specify shorter cutting limits on the order of 78:00 to 79:38 (check with the plant).

Standalone CD Recorders

Standalone CD recorders should not be used to make CDRs for replication. There is no provision for Index 0, and the location of Index 1 (the track mark) can only be as accurate as a manual button push. Also, when recording one track at a time, these standalone recorders work in Track-At-Once mode, which puts an E32 error onto the disk wherever the laser stops recording. Computer-based machines should be set to work in Disc-At-Once mode, which means that the CD must be written in one continuous pass.

Hidden Tracks

Find the hidden track is a little game the producer plays with the record-buyer. To make this possible, the mastering engineer can hide a track by inserting many short, blank 4-second "dummy tracks" at the end of the CD prior to the "hidden" one, which forces the listener to cue many times before he can reach it. Another method is to put several track marks within the "hidden" song, which causes ripping programs to break it up into pieces. Yet another way to hide a track is to have a track mark with no music for a minute or more.

Some CD players have the ability to rewind in front of track one; this is called the pregap or first Index 0. One company claimed to have the rights to putting hidden tracks in that position, but it's not permitted by the Red Book standard, and many plants will not press CDs with a hidden track in the pregap. To the best of my knowledge, there is no way to produce a Disc Description Protocol (DDP) master with this feature, so only CDR masters can be produced if the DAW allows hidden pregap tracks. DDPs are data files with the finished master encoded upon them. Since data discs contain very high levels of redundancy, the potential for errors is very small. These can be burned onto DVD-Rs or onto a CDR if there is space (the extra redundancy adds data that might prevent it from fitting on a CDR).

EDITING

I love editing, because it can generate a hundred smiles in a day! I think a whole book should be written on digital audio editing techniques, but ultimately the skill of fine editing can only be learned through guided experience: the school of hard knocks and an apprenticeship. The purpose of this short section is to discuss some of what is possible in editing. Using sophisticated workstations, we can perform edits that were impossible in the days of analog tape and the razor blade. I once spent 30 hours painstakingly editing a spoken-word version of a novel, a task which can now be accomplished in a single day. SADiE's playlist-editing mode, which allows us to "spill" virtual tape, makes this extremely easy.

The Tale of the Head and Tail

Editing heads (starts) and tails (ends) is an important skill based on musical knowledge but developed through experience. Because mechanical artifacts can easily distract the listener's attention from the emotional aspects of the music, a mastered work should generally be consistent and smooth. Consider the fade-up at the beginning of a song. When the music requires it, this fade-up can be made quite sharp (equivalent to a 90° cut). But a fast fade-up often sounds wrong with soft music, especially pieces that begin with solo vocal or acoustic instruments. A delicate acoustic guitar solo can sound abrupt if the noise of the room and the preamp is not brought up from silence at the right speed and timing.

NATURAL ANTICIPATION

We also have to be aware of the important role played by moments of natural anticipation: the human breath before the vocal; the movement of the guitarist's hand before a strum; or the movement of the fingers and keys prior to hearing a piano downbeat. Often it sounds unnatural to cut these off, making the opening appear choked. If the breath is better included, but sounds a bit loud, then a gentle fade-up can produce just the right result. I advise mixing engineers not to cut off the tops when sending songs for mastering, for the mastering engineer probably has better tools to fix these, and a quiet, meditative environment to make these artistic decisions properly.

TAIL NOISE CLEANUP

Sometimes the tail end of a song contains noise from musicians or equipment, which draws attention to itself by the transition from noise to the silence of the gap. Tools such as Cedar Retouch and Algorithmix Renovator are very useful in cleaning such noises and creating clean decays. Another common solution is called a follow fade, which is usually a cosine- or S-shaped fade to silence. A good mastering engineer may spend a minute or more on such a fade to ensure that the tail ambience or reverberation is not cut off, as the hiss or noise is brought invisibly to silence. We can take advantage of the fact that noise is masked by signal of the right amplitude, so the follow fade can and should be slightly slower than the natural decay. The delicate decay of a piano chord at the end of a tune should sound natural, even as we manipulate the fade-out to avoid or soften the thump of the release of the pedal. Fine editing can allow us to raise the gain at the tail, after having previously lowered it, in order to hear some inner detail.

FADE-OUTS

I think a good-sounding musical fade-out is one that makes us think the music is still going on; we're still tapping our feet even after the sound has ceased. Although we can apply the same cosine shape we use for tails, fade-outs are a distinct art in themselves. Typically, a fade-out will start slowly, and then taper off rapidly, mimicking the natural hand movement on a fader—there's nothing more annoying than a fade-out that lingers beyond its artistic optimum. On the other hand, a fade-out should not sound like it fell off a cliff, and often in mastering we get material that has to be repaired because the mix engineer dropped the tail of the fade too fast. Since editing is like whittling soap, I recommend that mix engineers send unfaded material so it can be refined in the mastering. It is difficult to satisfactorily repair a fade that was too fast at the end; sometimes an S-shape helps and sometimes we can apply a taper on top of the original slope.

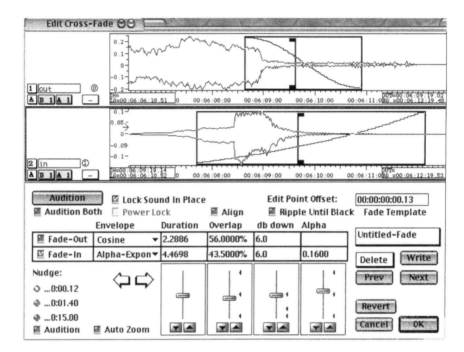

FIGURE 7.3
Adding a tail via a crossfade to artificial reverb.

ADDING TAILS

Although editing fades is like whittling soap, sometimes we're called upon to make more soap. If the musicians or instruments make a distracting noise during the ambient decay, the ambience will sound cheated or cut off if we perform a follow fade to remove the noises. Figure 7.3 is a fade-out to the right of which you can see the noise made by the musicians. Unfortunately, these noises occurred during the reverberant tail, so the ambience sounds cut off. The trick is to feed just the tail of the music into a high-quality artificial reverb and capture that in the workstation, which you can see in the bottom panel. Since the predelay of the reverb postpones its onset, its position can be adjusted in the DAW's crossfade window which allows us to carefully shape, time and level the transition to this artificial reverb in a manner that can sound completely seamless. Thus, we have performed the impossible: putting the soap back on the sculpture!

Sometimes an analog tape may have a lot of echoey print through or hiss noticeable at the tail of the tune. To add a new artificial tail can help, or editing to a digital safety version of the mix if it exists (so I advise clients to send me both versions).

Adding Room Tone

Room tone is essential between tracks of much natural acoustic and classical music. Recording engineers should bring samples of room tone to an editing session. Room tone is usually not necessary for pop productions, but if a recording gets very soft and you can hear the noise of the room, going sharply to audio black can be

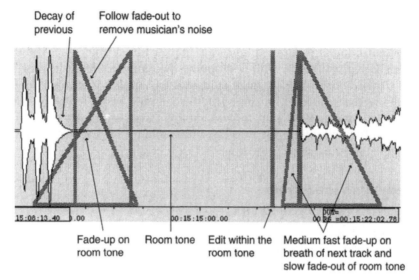

Decay of previous

Follow fade-out to remove musician's noise

Fade-up on room tone

Room tone

Edit within the room tone

Medium fast fade-up on breath of next track and slow fade-out of room tone

FIGURE 7.4
Editing room tone in an acoustic work requires considerable artistry. An edit must not call attention to itself.

disconcerting. The object is not to draw the listener's attention to the removal of noise as illustrated in Figure 7.4.

Room tone should be recorded in advance as a separate "silent take" with no musicians in the room. If this is not supplied (preferably 10 seconds or more), it is almost impossible for us to manufacture a convincing transition and we have to be satisfied with a fade to/from silence. Retouch or Renovator are handy at editing out glitches in room tone.

REPAIRING BAD EDITS

One type of bad edit is where the reverberation of one section has been cut off by the insertion of a new one. This used to happen in analog tape editing because we could not do intricate crossfades with a razor blade. But the error still occurs in classical music if a highly inexperienced producer instructed the musicians to begin the retake exactly at the intended edit point, instead of a few bars earlier which not only gives the musicians a running start and allows for a better music flow, but also generates the reverberant decay of the preceding note for the editor to work with. If the producer did not record the reverberation, the ear notices the cutoff of the reverb, which is not masked by the transient attack of the next downbeat. Luckily, when it comes to mastering, we can repair some of these bad edits even if the original takes are not available. The trick is to take apart the original edit, then feed an artificial reverb chamber to re-create the missing tail, then join it all back together.

Editing and assembling concert albums can be a great pleasure because they are the perfect example of suspension of disbelief. To edit applause requires a familiarity with natural applause, but real-life applause is almost never as short as 15 or 20 seconds, and real-life artists have to stop to tune their instruments. The object is to capture just the essence of the concert so that the home listener is never bored on replay. Cutting applause and ambience between different performances exercises the power of the workstation's crossfades. There must always be some degree of room tone (audience ambience) between numbers but the audience at the end of a loud number is more enthusiastic than before a quiet one. The transition will not sound realistic by simply dropping the ambience level; it must be done with a crossfade from loud to quiet ambience. My approach is to do the major cutting on one pair of tracks (for stereo), and wherever it needs transitional help, mix in a bed of compensating ambience on another track pair. I once put an audience loop under the only studio cut on a live album and to this day no one has been able to figure out which track is the ringer!

Leveling the Album

CONTEXT-BASED LEVELING

A piece of music which begins softly but follows one that ends loudly creates a potential problem. We may have to raise these beginnings, but the same soft level could be perfectly acceptable in the middle of a different piece. Similarly, a loud attack is amplified by the ear if it is preceded by silence. This is why albums and singles have to be leveled differently; the ballads often must be raised for single release.

The greater a recording's dynamic range, the harder it is to judge "average level" and you have to listen in several spots. I usually start with the loudest song on the album and find its highest point. I then engineer the processing to create the impact I'm looking for, hold the monitor at the predetermined gain, and make the rest of the songs work together at that monitor gain. This practice also helps prevent overprocessing or overcompression.[7] The rest of the album falls in line once the loudest song has its proper level and impact. During the processing of this loudest song, it's important to ensure an optimal gain structure in the chain of processors; this is the test for the rest of the album. Leveling and dynamics processing are inseparable. This is due to the fact that the output (makeup gain) of the processors also determines the song's loudness compared to the others. A more compressed song may sound louder than another even if its peaks don't hit full scale (0 dBFS). If you change the processing, you have also changed its level, so it all has to be done by ear. After working on the loudest song and saving the settings, I usually go to the first song and work in sequence. Then the second song and next I check the transition between the first and second. In a good mastering room, this transition will usually work without any fine-tuning because we've been monitoring at a consistent gain. If one song appears too loud or soft in context, I make a slight adjustment in level until they work together, or sometimes increase the spacing to "clear the ear." So you can see why it's important to have the album in proper order before mastering!

Everything Louder Than Everything Else

After leveling and processing the last song, I always review song numbers one and two, to make sure they still fit well into the context as there may be a tweak that

---

[7]Mix engineers follow a similar practice, beginning the mix from the loudest point of the song.

can further optimize them. Or, I might find that the album has been growing in amplitude due to ear fatigue and the latter songs may need to be lowered.

Overzealous leveling practice can produce a *Domino Effect*. Suddenly, the song which used to be the loudest, doesn't sound as loud as it did before. This is psycho-acoustics at work—not every song can be the loudest! If the loudest song was good enough before, the problem may be the unintentional escalation. Instead of trying to push the loudest song further, thereby squashing it with the limiter, I try to lower the previous song by even a few tenths of a decibel, which will help restore the impact of the next song by use of contrast.

FURTHER READING

This edited chapter is from Katz, B. (2007). *Mastering Audio: The Art and the Science* 2nd Edition, Focal Press.

SECTION C
To Market

Chapter 8

Product Manufacture: How the Record Is Made

David Miles Huber and Robert Runstein

In This Chapter

189

In this chapter, David Huber and Robert Runstein introduce how our music is transferred to a distributable physical medium such as a CD or a record.

INTRODUCTION

One of the greatest misconceptions surrounding the music, visual, and other media-related industries is the idea that once you walk out the door of a studio with your final master in hand, the creative process of producing a project is finally over. All that you have left to do is hand the file, CD, or other medium over to a duplication facility and—Ta-Dah!—the buying public will be clamoring for your product—website, and merchandise. Obviously, this scenario is almost always far from the truth. Now that you have the program content in hand, you have to think through and implement your master plan, if your product is to make it into the hands of the consumer.

I have a 1st rule of recording … that there are no rules, only guidelines. This actually isn't true. There is one rule here: If you don't pre-plan and follow through with these plans once the project is recorded, you can be fairly sure that your project will sit on a shelf, or worse, you'll have 1000 CDs sitting in your basement that'll never be heard—a huge shame given the blood, sweat 'n' tears that went into making it.

Once the recording and mixing phases of a project have been completed (assuming that you've done your homework as to your audience, distribution methods, live and Web marketing presence, production budgeting, etc.), the next step toward getting the product out to the people is to transform the completed song or project into a form that can be mass-produced, distributed, marketed, and "SOLD."

Given the various technologies that are available today, this could take the form of a CD, DVD, CD-ROM, vinyl record, or encoded file. Each of these media types has its own set of manufacturing and distribution needs that require a great deal of careful attention throughout each step of the manufacturing and/or creation process.

CHOOSING THE RIGHT FACILITY AND MANUFACTURER

Just as recording studios have their own unique personalities and particular "sound," the right mastering and duplication facilities may also have a profound effect on the outcome of a project. If a project is being underwritten and distributed by an independent or major record label, they will generally be fully aware of their production needs and will certainly have an established production and manufacturing network in place. If, however, you are distributing the project yourself, the duty of choosing the best facility or manufacturing organization that'll fit your budget and quality needs is all yours.

A number of resources exist that can help you find such manufacturers. For starters, Billboard Online (www.billboard.com) provides numerous services for searching out media mastering and manufacturing facilities in the USA. They offer resource magazines (such as the *Billboard Tapel Disc Directory*), as well as a free online search database for *Billboard* magazine subscribers. *The Mix Master Directory* (Intertec Publishing, 6400 Hollis St., Suite 12, Emeryville, CA 94608; 1–510–653–3307, www.mixonline.com) publishes an annual directory of industry-related products and services. This directory (which is sent out as a supplement to the January

Mix issue) provides a comprehensive listing of manufacturers, recording studios, producers, engineers, music business services, and so on, and is cross-referenced by product and service categories. The *Recording Industry Source Book* (http://artist-pro.com, 447 Georgia Street, Vallejo, CA 94590; 1–707–554–1935, www.isource-book.com or www.artistpro.com) also includes a full listing for these companies. Another simple but effective resource is to look at the back page ads in most music- and audio-related magazines.

In the UK, there are a number of sources for similar information. One of the major sources is *The White Book* (www.whitebook.co.uk). As previously mentioned the trade magazines are an excellent source of information such as MusicWeek, ProSound News, Resolution, Sound On Sound, and MusicTech Magazine, etc., which carry some listings and advertisements that might prove useful.

Manufacturing facilities come in two types: those that perform and offer all of their services "in-house" (on the premises), and those that "out-source" (contract with other business or individuals to perform various services). Neither of these types is good or bad. On one hand, in-house facilities are able to handle all of the phases of producing a finished product, from beginning to end; these facilities are often large and expensive to equip (meaning that one may not be located nearby). On the other hand, manufacturers and duplicators that farm out projects may not have total control over their production timeline, but are often able to offer personalized, one-on-one service.

As with any part of the production process, it's always wise to do a full background check on a production facility and even compare prices and services from at least three manufacturing houses. A good way to check out a place is to ask for a promotional pack (which includes product and art samples, facility, service options, and a price sheet). You might want to ask about former customers and their contact information (so you can e-mail them about their experiences). Most important, once you've chosen a mastering and/or manufacturing facility, it's extremely important that you be given art proofs and test pressings BEFORE the final products are mass duplicated.

Making a test pressing is well worth the time and money; the alternative is to receive a few thousand copies at your doorstep, only to find that they're not what you wanted—a far more expensive, frustrating, and time-consuming option. It's never a good idea to assume that a manufacturing or duplication process is perfect and doesn't make mistakes. Remember, Murphy's Law can pop up at any time!

The remainder of this chapter we will largely concentrate on the manufacturing process for the various mass-market media. At the end of this chapter, we'll briefly discuss highlights of what is likely the most important part of the production process.

CD MANUFACTURING

Although a project can wind up in any number of final medium forms, as of this writing, the compact disc (CD) is still the easiest and most widely recognized medium for distributing physical music, although with the advent of iTunes®, digital downloading is on a fast increase. These 4¾" silvery CDs (shown in Figure 8.1) contain digitally encoded information (in the form of microscopic pits) that's capable of yielding playing times of up to 74 minutes, at a standard sampling rate of 44.1 kHz.

A CD pit is approximately half a micrometer wide, and a standard manufactured disc can hold about 2 billion pits. These pits are encoded onto the disc's surface in a spiraling fashion, similar to that of a record's grooves, except that 60 CD spirals can fit in the groove of a single long-playing record. The CD spirals also differ from a record in that they travel outward from the center of the disc and are impressed into the plastic substrate, which is covered with a thin coating of aluminum (or occasionally gold) so that the laser light can be reflected back to a receiver. When the disc is placed in a CD player, a low-level infrared laser is alternately reflected and not reflected back to a photosensitive pickup. In this way, the reflected data is modulated so that each pit edge represents a binary 1, and the absence of a pit edge represents a binary 0 as shown in Figure 8.2. Upon playback, the data is then demodulated and converted back into an analog form.

FIGURE 8.1
The compact disc. (Courtesy of 51 bpm.com, www.51bpm.com.)

Songs or other types of audio material can be grouped on a CD as indexed "tracks." This is done via a subcode channel lookup table, which makes it possible for the player to identify and quickly locate tracks with frame accuracy.

FIGURE 8.2
Transitions between a pit edge (binary 1) and the absence of a pit edge (binary 0).

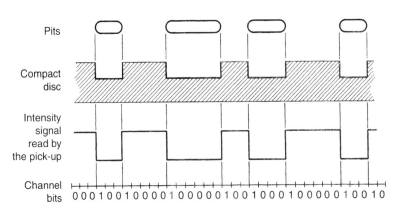

©2009 David Miles Huber and Robert Runstein

Subcodes are event pointers that tell the player how many selections are on the disc and where their beginning address points are located. At present, eight subcode channels are available on the CD format, although only two (the P and Q subcodes) are used.

The Process

In order to translate the raw PCM of a music or audio project into a format that can be understood by a CD player, a CD mastering system must be used. Modern mastering systems often come in two flavors: those that are used by professional-mastering facilities, and CD-R/CD-RW hardware/software systems that allow a personal computer to easily and cost-effectively burn CDs.

As discussed in Chapter 6, both system types allow audio to be entered into the system, assemble tracks into the proper order or sequence, and enter proper gap times between tracks (in the form of index timings). Depending on the system, cuts might also be processed using crossfades, volume, EQ, and other parameters. Once assembled, the project can be "finalized" into a media form that can be directly accepted by a CD manufacturing facility.

In the case of a professional system, this media could take the form of either a DVD-R or an Exabyte data tape containing a Disc Description Protocol (DDP) file, 3/4" U-matic videotape (using a Sony PCM-1630 type digital processor) or CD-Recordable (CD-R). Although digitally encoded tapes and DDP adorned DVD-Rs are considered to be the most reliable, the equipment that's required to prepare a master is extremely expensive. This is the reason why the majority of final CD masters being received by CD pressing plants are user-created CD-R discs.

Once the manufacturing plant has received the recorded media, the next stage in the process is to cut the original CD master disc. The heart of such a CD cutting system is an optical transport assembly that contains all the optics necessary to write the digital data onto a reusable glass master disc that has been prepared with a photosensitive material.

After the glass master has been exposed using a special recording laser, it's placed in a developing machine that etches away the exposed areas to create a finished master. An alternative process, known as non-photoresist, etches directly into the photosensitive substrate of the glass master without the need for a development process.

After the glass or CD master disc has been cut, the compact disc manufacturing process can begin. This process is illustrated in Figure 8.3. Under extreme clean room conditions, the glass disc is electroplated with a thin layer of electro-conductive metal. From this, the negative metal master is used to create a "metal mother," which is used to replicate a number of metal "stampers" (metal plates which contain a negative image of the CD's data surface). The resulting stampers make it possible for machines to replicate clear plastic discs that contain the positive encoded pits, which are then coated with a thin layer of foil (for increased reflectivity) and encased in clear resin for stability and protection. Once this is done, all that

FIGURE 8.3
Various phases of the CD manufacturing process: (a) the mastering studio, where the process begins; (b) once the graphics are approved, the project's packaging can move onto the printing phase; (c) while the packaging is being printed, the approved master can be burned onto a glass master disc; (continues)

(a)

(b)

(c)

(d)

FIGURE 8.3
(continued) (d) next, the master stamper (or stampers) is placed onto the production line for CD pressing; (e) the freshly stamped discs are cooled and checked for data integrity; (f) labels are then silk screened printed onto the CDs; (continues)

(e)

(f)

FIGURE 8.3
(continued) (g) finally, the printed CDs are checked before being inserted into their finished packaging. (Courtesy of Disc Makers, Inc., www. discmakers.com.)

(g)

FIGURE 8.4
CDR-830 "Burnit" Compact Disc Recorder. (Courtesy of HHB Communications Ltd., www.hhb.co.uk.)

remains is the screen printing process and final packaging. The rest is in the hands of the record company, the distributors, marketing, and you.

CD BURNING

Before the proliferation of CD-recording hardware and software (shown in Figures 8.4–8.6), the only way to hear how your final CD would sound was to press a "one-off" disc. This meant that the CD manufacturer had to go through the aforementioned process of creating a glass master and "cut" a single or limited run of CDs for the producer, artist, and/or record company as a reference disc. As you could guess, this made it a time-consuming and expensive process that was only available to companies and individuals with big bucks.

Nowadays, the process of burning a CD on a Mac or PC has become so widespread and straightforward that producers, engineers, artists, and the general public use it as the preferred medium for creating and distributing music.

Although most manufacturing plants receive master CDs that have been burned onto a CD-R, it's interesting to note that many of these discs don't pass the basic requirements that have been set forth for creating an acceptable Red Book Audio

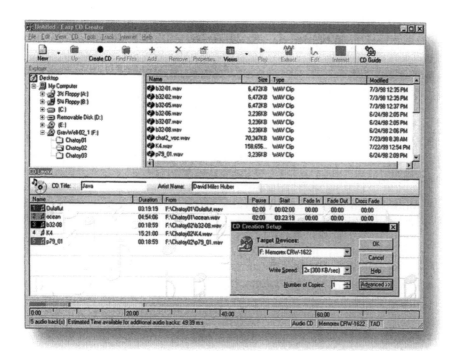

FIGURE 8.5
Easy CD Creator CD
burning software.
(Courtesy of Roxio,
www.roxio.com.)

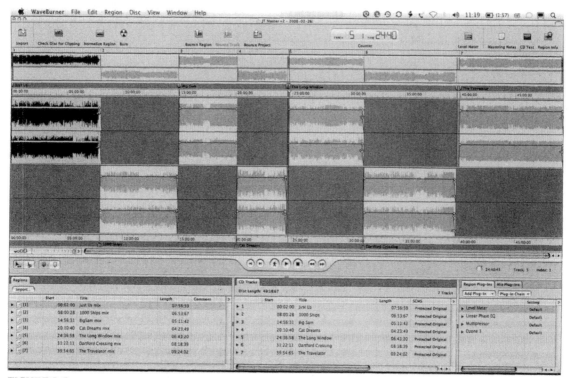

FIGURE 8.6
Waveburner by Apple. (Courtesy of Apple.)

©2009 David Miles Huber and Robert Runstein

CD (the standard industry specification). Some of the problems associated with CD-Rs that have been burned on a desktop system include the following:

- *Excessive data errors*: This can lead to mass-produced CDs that have problems when being played on older or less reliable CD players. These errors could crop up because of such factors as hardware/software reliability problems or media integrity.
- *Discs that haven't been "closed"*: It's very important that the master disc be closed (a coding process that ensures that no other sessions or data can be added to the disc). Most CD-mastering software packages will give you the option of closing or "finalizing" the disc upon burning.
- *Multi-session discs*: Final master discs should never contain multiple sessions (in which music cuts or program material is added at a later time to an existing CD-R). The disc should be recorded and finalized in the "disc-at-once" mode (meaning the disc was burned from beginning to end, without any interruptions in the laser burning process).
- *Inaccurate index marker points*: Index markers tell the CD player where the tracks begin and end on a disc. If the markers are wrong, the program could begin early or cut off parts of a song. Once a disc has been cut, always listen to a disc to check for accurate index markers.

In fact, once you've checked the beginning and end marker points, it's always wise to critically listen to the disc from beginning to end. Never forget that Murphy's Law lurks around every corner! Once you've agreed that the CD sounds great, it's always a good idea to burn an extra master that can be set aside for safekeeping, in case something happens to the original production master.

It's interesting to note that many CD burning programs will allow information (such as title, artist name/copyright, and track name field code info) to be written directly into the CD's subcode area. This is often a good idea, as important artist, copyright, and track identifiers can be directly embedded within the CD itself. As a result, illegal copies will still contain the proper copyright and artist info, and discs that are loaded into a computer or media player will often display these fields.

Currently, two types of CD-recording media are commonly found: the CD-R and CD-RW (rewritable) disc. These media use a dye whose reflectivity can be altered, so that data can be burned to disc using a number of available writing options:

- *Disc-at-once*: This mode continuously writes the data onto CD without any interruptions. All of the information is transferred from hard disc to the CD in a single pass, with the lead-in, program data, and lead-out areas being written to disc as an uninterrupted event.
- *Track-at-once*: This allows a session to be written as a number of discrete events (called tracks). With the help of special software, the disc can be read before the final session is fixated (a process that "closes" the disc into a final form that can be read by any CD or CD-ROM drive).
- *Multi-session*: Discs written in this mode allow several sessions to be recorded onto a disc (each containing its own lead-in, program data, and lead-out

areas), thereby allowing data to be recorded onto the free space of a previously recorded CD. It should be noted that older drives might not be able to read this mode and will only read the first available session.

Although the altering of the data pits on a CD-R is permanent, CD-RW can be erased and rewritten any number of times (often figured in the thousands). A specially designed CD-RW drive (which can also burn standard CD-Rs), is excellent for creating data backups and media archiving. In addition, many of the newer CD and MP3 disc players are capable of reading CD-RW media.

Rolling Your Own

With the rise of Internet audio distribution and the slow but steady breakdown of the traditional record company distribution system, bands, and individual artists have begun to produce, market, and sell their own music on an ever-increasing scale. This age-old concept of the "grower" selling directly to the consumer is as old as the town square produce market. However, by using the global Internet economy, independent distribution, fanzines, live concert sales, etc., savvy independent artists are taking matters into their own hands (or are smart enough to combine with the talents of others) by learning the inner workings of the music business. In short, artists are taking business matters more seriously to reap the fruits of their labor and craft ... something that has never been and never will be an easy task.

Beyond the huge tasks of marketing, gigging, and general business practices, many musicians are also taking on the task of burning, printing, packaging, and distributing their own CDs from the home or business workplace. This homespun strategy allows for small runs to be made in an "on-demand" basis, without tying up financial resources and storage space in CD inventories.

Creating a system for burning CD-Rs for distribution can range from being a simple home computer setup that creates discs on an individual basis to sophisticated replication systems that can print and burn stacks of CD-Rs or DVD-Rs under robot control at the simple touch of a button as shown in Figure 8.7.

FIGURE 8.7
Elite Pro CD burning and printing system. (Courtesy of Disc Makers, Inc., www. discmakers.com.)

Burning Speeds

Whenever you see the specs on a CD-R or CD-RW burner that look like $32 \times 10 \times 40$ (three numbers that are separated by the character "$\times$"), the numbers indicate the various read and writing speeds of the CD drive. The "$\times$" stands for the device's data transfer speed as multiples of 150 kB/second. The first number (32 in the above example) indicates the speed that the drive is capable of writing data onto a CD-R disc. In the above example, the drive can write at transfer speeds of up to 32×150 kB/sec = 4800 kB/sec The second number represents the speed that the drive can rewrite data onto a CD-RW disc (i.e., 10×150 kB/sec = 1500 kB/sec). The final number indicates the top speed that the drive can read at (i.e., 40×150 kB/sec = 6000 kB/sec).

Finding the optimum CD-R burning speed for your computer or replicator is a topic that's best left as a debate between buddies over a pint of beer. There are those that passionately feel that burning at lower speeds will improve the burning process due to improved disc stability and optimum laser performance, while others will argue that newer media dyes, improved laser assemblies, and numerous amounts of published data will prove them wrong—that performance actually improves at higher (though not always maximum) writing speeds. I will bow out of this debate by challenging you to research the data, the articles, and the many message postings that have been dedicated to this subject.

CD Labeling

Once you've burned your own CD-R/RW or DVD-R/RW, there are a number of options for printing labels onto the newly burned discs (burning the disc first will often reduce data errors that can be introduced by dust, fingerprints, or scratches due to handling):

FIGURE 8.8
Neato CD Labeler kit.
(Courtesy of Neato
LLC, www.neato.com.)

- *Using a felt-tip pen*: This is the easiest and fastest way to label a disc. However, water-based ink pens should be used, because permanent markers use a solvent that can permeate the disc surface and cause damage to either the reflective or dye layer. When properly done, this is an excellent option for archived discs.
- *Label printing kits*: "Stick-on" labels (as shown in Figure 8.8) that have been printed using specially designed software and an inkjet or laser printer are one of the least expensive options. Although their

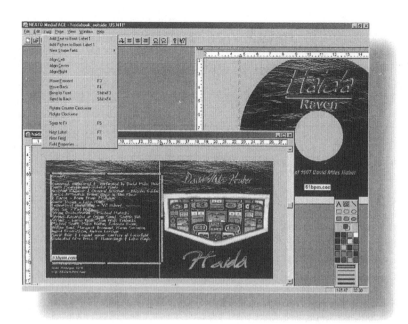

FIGURE 8.9
Mediaface label
printing program.
(Courtesy of Neato
LLC, www.neato.com.)

design has improved over the years, you should be aware that some adhesives could peel off, leak over time, or contain solvents that might adversely affect the disc. This professional-looking approach is often excellent for use on non-archival products.

- **CD printers**: Specially designed inkjet or laser printers are able to print high-quality, full-color layouts onto the face of a printable (white- or silver-faced) disc. This is a cost-effective option for those who burn discs in small batch runs and want a professional look and feel.

Although stand-alone programs are available, most of the above-mentioned printing kits and CD printers include a label printing program for creating and printing professional-looking CDs, CD books, and trays (as well as labels for DATs, cassettes, and almost any other medium you can think of). These programs let you import graphics and position text to create and print out personalized, professional-looking labels as shown in Figure 8.9. In addition to these programs, word processing templates are often available (most often for Microsoft Word) that let you import graphics and position text as a document that can be opened and printed directly from the word processor, without the need for a special program.

DVD BURNING

Of course, on a basic level, DVD burning technology has matured enough to be available and affordable to the general Mac and PC public.

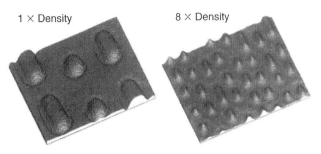

1 × Density   8 × Density

FIGURE 8.10
Detailed relief
showing standard CD
and DVD pit densities.

From a technical standpoint, these CD-drive compatible discs differ from the standard CD format in several ways. The most basic of these are:

- An increased data density due to a reduction in pit size (as shown in Figure 8.10.)
- Double-layer capabilities (due to the laser's ability to focus on two layers of a single side)
- Double-side capabilities (which again doubles the available data size).

In addition to the obvious benefits that can be gained from increasing the data density of a standard CD from 650 MB to a maximum of 17 GB, DVD discs allow for much higher data transfer rates, making DVD the ideal medium for the following applications:

- The simultaneous decoding of digital video and surround-sound audio
- Multi-channel surround sound
- Data- and access-intensive video games
- High-density data storage.

Using extensive, high-quality compression techniques, this technology has breathed new life into the home entertainment industries, allowing computer fans to have access to increased game and multimedia storage capacity and home viewers to enjoy master-quality audio and video programs in a digital surround-sound environment.

As DVD-ROM and writable drives have become commonplace, affordable data backup and mastering software has come onto the market that brought the art of DVD mastering to the masses. Even high-level DVD production is now possible in a desktop environment, although creating a finished product for the mass markets is often an art that's often best left to professionals who are familiar with the finer points of this complex technology.

CD AND DVD HANDLING AND CARE

Here are a few basic handling tips for CDs and DVDs (including the recordable versions) from the National Institute of Standards and Technology:

DO

- Handle the disc by the outer edge or center hole (your fingerprints may be acidic enough to damage the disc).
- Use a felt-tip permanent marker to mark the label side of the disc. The marker should be water or alcohol based. In general, these will be labeled "non-toxic." Stronger solvents may eat though the thin protective layer to the data.

- Keep discs clean. Wipe with a cotton fabric in a straight line from the center of the disc toward the outer edge. If you wipe in a circle, any scratches may follow the disc tracks, rendering them unreadable. Use a CD/DVD-cleaning detergent or isopropyl alcohol to remove stubborn dirt.
- Return discs to their cases immediately after use.
- Store discs upright (book style) in their cases.
- Open a recordable disc package only when you are ready to record.
- Check the disc surface before recording.

DON'T

- Touch the surface of a disc.
- Bend the disc (as this may cause the layers to separate).
- Use adhesive labels (as they can unbalance or warp the disc).
- Expose discs to extreme heat or high humidity, for example, don't leave them in sun-warmed cars.
- Expose discs to extreme rapid temperature or humidity changes.
- Expose recordable disc to prolonged sunlight or other sources of ultraviolet light.

ESPECIALLY DON'T

- Scratch the label side of the disc (it's often more sensitive than the transparent side).
- Use a pen, pencil, or fine-tipped marker to write on the disc.
- Try to peel off or reposition a label (it could destroy the reflective layer or unbalance the disc).

VINYL DISC MANUFACTURE

Although the popularity of vinyl has waned in recent years (as a result, of course, of the increased marketing, distribution, and public acceptance of the CD), the vinyl record is far from dead. In fact, for consumers that range from Dance DJ hip-hipsters to die-hard classical buffs, the record is still a viable medium. However, the truth remains that many record pressing facilities have gone out of business over the years, and there are far fewer mastering labs that are capable of cutting "master lacquers." It may take a bit longer to find a facility that fits your needs, budget, and quality standards, but it's definitely not a futile venture.

Disc Cutting

The first stage of production is the disc-cutting process. As the master is played from a digital source or on a specially designed tape playback machine, its signal output is fed through a disc-mastering console to a disc-cutting lathe. Here, the electrical signals are converted into the mechanical motions of a stylus and are cut into the surface of a lacquer-coated recording disc.

Unlike the CD, a record rotates at a constant angular velocity, such as 33⅓ or 45 revolutions per minute (rpm), and has a continuous spiral that gradually moves from the disc's outer edge to its center. The recorded time relationship can be reconstructed by playing the disc on a turntable that has the same constant angular velocity as the original disc cutter.

The system that's used for recording a stereo disc is the 45/45 system. The recording stylus cuts a groove into the disc surface at a 90° angle, so that each wall of the groove forms a 45° angle with respect to the vertical axis. Left-channel signals are cut into the inner wall of the groove and right-channel signals are cut into the outer wall, as shown in Figure 8.11. The stylus motion is phased so that L/R channels that are in-phase (a mono signal or a signal that's centered between the two channels) will produce a lateral groove motion (see Figure 8.12a), while out-of-phase signals (containing channel difference information) will produce a vertical motion that changes the groove's depth (see Figure 8.12b). Because mono information relies only on lateral groove modulation, an older disc that has been recorded in mono can be accurately reproduced with a stereo playback cartridge.

Disc-Cutting Lathe

The main components of a vinyl disc-cutting lathe are the turntable, lathe bed and sled, pitch/depth-control computer, and cutting head. Basically, the lathe illustrated

FIGURE 8.11
The 45/45-cutting system encodes stereo waveform signals into the grooves of a vinyl record.

FIGURE 8.12
Groove motion in stereo recording. (The solid line is the groove with no modulation.) (a) In-phase, (b) out-of-phase.

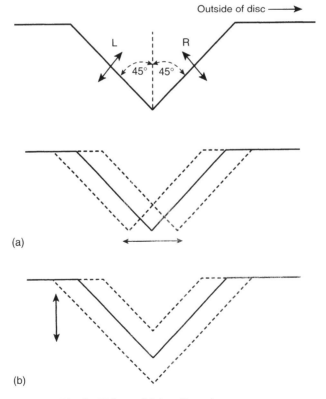

©2009 David Miles Huber and Robert Runstein

in Figure 8.13 consists of a heavy, shock-mounted steel base (A). A weighted turntable (B) is isolated from the base by an oil-filled coupling (C), which reduces wow and flutter to extremely low levels. The lathe bed (D) allows the cutter suspension (E) and the cutter head (F) to be driven by a screw feed that slowly moves the record mechanism along a sled in a motion that's perpendicular to the turntable.

Cutting Head

The cutting head translates the electrical signals that are applied to it into mechanical motion at the recording stylus. The stylus gradually moves in a straight line toward the disc's center hole as the turntable rotates, creating a spiral groove on the record's surface. This spiral motion is achieved by attaching the cutting head to a sled that runs on a spiral gear (known as the lead screw), which drives the sled in a straight track.

The stereo-cutting head shown in Figure 8.14 consists of a stylus that's mechanically connected to two drive coils and two feedback coils (which are mounted in a permanent magnetic field) and a stylus heating coil that's wrapped around the tip of the stylus. When a signal is applied to the drive coils, an alternating current flows through them creating a changing magnetic field that alternately attracts and repels the permanent magnet. Because the permanent magnet is fixed, the coils move in proportion to a field strength that causes the stylus to move in a plane that's 45° to the left or right of vertical (depending on which coil is being driven).

Pitch Control

The head speed determines the "pitch" of the recording and is measured by the number of grooves, or lines per inch (lpi), that are cut into the disc. As the head speed increases, the number of lpi will decrease, resulting in a corresponding decrease in playing time. Groove pitch can be changed by:

- Replacing the lead screw with one that has a finer or coarser spiral.
- Changing the gears that turn the lead screw, so as to alter the lead screw's rotation speed.

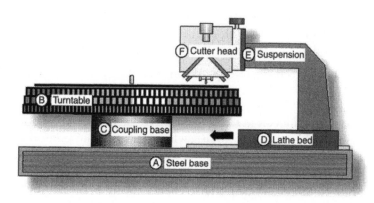

FIGURE 8.13
A disc-cutting lathe with automatic pitch and depth control.

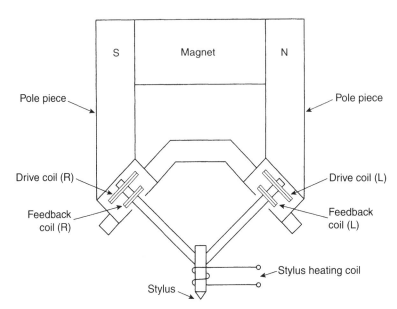

FIGURE 8.14
Simplified drawing of a stereo-cutting head.

- Vary the lead screw's rotation by changing the motor's speed (a common way to vary the program's pitch in real time).

The space between grooves is called the land. Modulated grooves produce a lateral motion that's proportional to the in-phase signals between the stereo channels. If the cutting pitch is too high (causing too many lpi, which closely spaces the grooves) and high-level signals are cut, it's possible for the groove to break through the wall into an adjacent groove (causing "over cut") or for the grooves to overlap (twinning). The former is likely to cause the record to skip when played, while the latter causes either distortion or a signal echo from the adjacent groove (due to wall deformations). Groove echo can occur even if the walls don't touch and is directly related to groove width, pitch, and level.

These cutting problems can be eliminated either by reducing the cutting level or by reducing the lines per inch. A conflict can arise here as a louder record will have a reduced playing time, but will also sound brighter, punchier, and more present (due to the Fletcher-Munson curve effect). Because record companies and producers are always concerned about the competitive levels of their discs relative to those that are cut by others, they're reluctant to reduce the overall cutting level.

The solution to these level problems is to continuously vary the pitch so as to cut more lines per inch during soft passages and fewer lines per inch during loud passages. This is done by splitting the program material into two paths: undelayed and delayed. The undelayed signal is routed to the lathe's pitch/depth-control computer (which determines the pitch needed for each program portion and varies the lathe's screw motor speed). The delayed signal (which is usually achieved by using a high-quality digital delay line) is fed to the cutter head, thereby giving the pitch/depth control computer enough time to change the lpi to the appropriate pitch.

Recording Discs

The recording medium used on the lathe is a flat aluminum disc that's coated with a film of lacquer, which is dried under controlled temperatures, coated with a second film, and then dried again. The quality of these discs (called lacquers) is determined by the flatness and smoothness of the aluminum base. Any irregularities in this surface (such as holes or bumps) will cause similar defects in the lacquer coating. Lacquers are always larger in diameter than the final record, which makes it easy to handle them without damaging the grooves. For example, a 12-inch album is cut on a 15-inch lacquer and a 7-inch single is cut on a 10- or 12-inch lacquer. As always, it's wise to cut a reference test lacquer in order to hear how the recording will sound after being transferred to disc.

The Vinyl Mastering Process

Once the mastering engineer sets a basic pitch on the lathe, a lacquer is placed on the turntable and compressed air is used to blow any accumulated dust off the lacquer surface. A chip suction vacuum is started and a test cut is made on the outside of the disc to check for groove depth and stylus heat. Once the start button is pressed, the lathe moves into the starting diameter, lowers the cutting head onto the disc, starts the spiral and lead-in cuts, and begins playing the master production tape. As the side is cut, the engineer can fine-tune any changes to the previously determined console settings. Whenever an analog tape machine is used, a photocell mounted on the deck senses white leader tape between the selections on the master tape and signals the lathe to automatically expand the grooves to produce track bands. After the last selection on the side, the lathe cuts the lead-out groove and lifts the cutter head off the lacquer.

This master lacquer is never played, as the pressure of the playback stylus would damage the recorded soundtrack (in the form of high-frequency losses and increased noise). Reference lacquers (also called reference acetates or simply acetates) are cut to hear how the master lacquer will sound.

After the reference is approved, the record company assigns each side of the disc a master (or matrix) number that the cutting room engineer scribes between the grooves of the lacquer's ending spiral. This number identifies the lacquer in order to eliminate any need to play the record and often carries the mastering engineer's personal identity mark. If a disc is remastered for any reason, some record companies retain the same master numbers; others add a suffix to the new master to differentiate it from the previous "cut."

When the final master arrives at the plating plant, it is washed to remove any dust particles and is then electroplated with nickel. Once the electroplating is complete, the nickel plate is pulled away from the lacquer. If something goes wrong at this point, the master will be damaged, and the master lacquer must be re-cut.

Vinyl Disc Plating and Pressing

The nickel plate that's pulled off the master (called the matrix) is a negative image of the master lacquer shown in Figure 8.15. This negative image is then

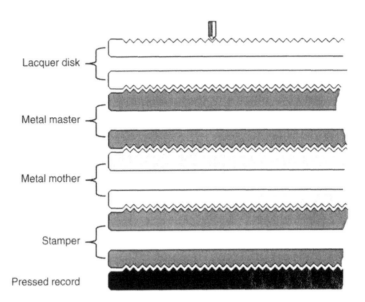

FIGURE 8.15
The various stages in the plating and pressing process.

Lacquer disk

Metal master

Metal mother

Stamper

Pressed record

electroplated to produce a nickel positive image called a mother. Because the nickel is stronger than the lacquer disc, several mothers can be made from a single matrix. Since the mother is a positive image, it can be played as a test for noise, skips, and other defects. If it's accepted, the mother can be electroplated several times, producing stampers that are a negative images of the disc (a final plating stage that's used to press the record).

The stampers for the two sides of the record are mounted on the top and bottom plates of a hydraulic press. A lump of vinylite compound (called a biscuit) is placed in the press between the labels for the two sides. The press is then closed and heated by steam to make the vinylite flow around the raised grooves of the stampers. The resulting pressed record is too soft to handle when hot, so cold water is circulated through the press to cool it before the pressure is released. When the press opens, the operator pulls the record off the mold and the excess (called flash) is trimmed off after the disc is removed from the press. Once done, the disc's edge is buffed smooth and the product is ready for packaging, distribution, and sales.

PRODUCING FOR THE WEB

In this day of surfing and streaming media from the Internet, it almost goes without saying that the World Wide Web (WWW) has become an important marketing tool for cost-effectively getting downloadable songs, promotional materials, touring info, and liner notes out to mass audiences. As with other media, mastering for the Internet can either be complicated, requiring professional knowledge and experience, or it can be a straightforward process that can be carried out from a desktop computer. It's a matter of meeting the level of professionalism and development that's required by the site.

In this iPod® world of MP3s, AAC (MP4s) Windows Media, desktop video, Internet radio stations, and who knows what other types of streaming media, the rule that all cyber-producers live by is bandwidth. Basically, the bandwidth of a media and delivery/receiving system refers to the ability to squeeze as much data (often compressed data) through a wire, wireless, or optical pipeline in as short a time as is possible. Transmitting the highest audio and/or video feed over a limited bandwidth will often require specialized (and often accessible) production tools and is outside the remit of this book.

Beyond this, an even more important tool is mastery of the medium, mass marketing, and good eyes and ears for design layout and media management.

DVD video/audio format:

Format	Sample Rate (kHz)	Bit Rate	Bit(s)	Ch	Common Format	Compression
PCM	48, 96	16, 20, 24	Up to 6.144 mbps	1–8	48 kHz, 16 bit	None
AC3	48	16, 20, 24	64–448 kbps	1–6.1	192 kbps, stereo	AC3 and 384 kbps, 448 kbps
DTS	48, 96	16, 20, 24	64–1536 kbps	1–7.1	377 or 754 kbps for stereo and 754.5 or 1509.25 kbps for 5.1	DTS coherent acoustics
MPEG-2	48	16, 20	32–912 kbps	1–7.1	Seldom used	MPEG
MPEG-1	48	16, 20	384 kbps	2	Seldom used	MPEG
SDDS	48	16	Up to 1289 kbps	5.1, 7.1	Seldom used	ATRAC

FURTHER READING

This edited chapter is from Huber, D. & Runstein, R. (2005). *Modern Recording Techniques*. Focal Press, London.
Also see http://modrec.com

Chapter 9

Publicity and Radio Promotion

Paul Allen

In This Chapter

Once the music is in its final state from the factory, it is the time when a number of concurrent considerations come into play, namely the Publicity and Marketing of your music. In this chapter, Paul Allen discusses the typical publicity machine and radio promotion used behind most traditional label-based releases.

INTRODUCTION

A traditional recording contract, simply stated, says that the artist will create recordings and the label will market the recordings. An important element of marketing planning and execution by a label is promotion, and publicity is typically a part of that overall promotional effort by the label. Labels usually handle publicity for the artist's recording career and for news and press releases about the label itself. Sometimes, artists will hire a personal publicist to handle other areas of their life and career.

This chapter is designed to give an overview of the publicity department at a record label, including its responsibilities and how publicity contributes to the success of a recorded music project.

LABEL PUBLICITY

The objective of label publicity is to place non-paid promotional messages into the media on behalf of the artist's recorded music project. That can range from a small bullet point in *Rolling Stone* to an appearance on the "Late Show with David Letterman." Appearances in the traditional media and on the web contribute to the success of the label's promotional plan to put the artist in front of music fans to support the marketing of the label's music.

The theory is that the more positive impressions consumers receive about a recording, the more likely they are to seek additional information about the recording and to purchase it. Advertising planners use the term *reach and frequency* as they compile a strategy and its related budget. This means they plan an affordable ad campaign that can "reach" sufficient numbers of their target market with the "frequency" necessary for them to remember the message and act by purchasing. Publicity becomes a nice complement to that strategy without the direct costs of paid advertising.

Label publicity on behalf of a recording artist, like any positive publicity, has a certain credibility that paid advertising does not. After all, a journalist thought the artist was interesting enough to write an article about him or her or a review of the music, and a publisher thought it was interesting enough to make editorial space in a magazine, a website, or a newspaper to present the story or run the review.

Advertising is the handcrafted, paid message of the record company's marketing department designed to sell recordings. It is someone's crafted message with the intent of getting into consumer's wallets. However, an effective publicity campaign can create an interest by journalists and online entertainment bloggers to write articles about artists and their recordings in the not-so-commercial setting of a feature article. An article in a newspaper, magazine, or website can suggest to the consumer that there is something more to the label's artist than just selling commercial music. Published articles and TV magazine-style stories (for example, "60 Minutes") add credibility to the artist as an "artist" in a way that paid advertising cannot, and tend to be a "filter" of the important from the not-so-important.

There are key differences between publicity from the label and advertising placed by the label. Label publicists generally create and promote messages to the media that are informative in nature and do not have a hard "sell" to them. On the other hand, advertising is designed to influence and persuade the consumer to purchase CDs, videos, ringtones, and other ancillary products created by the label.

Publicity in the Music Business—A Historical Perspective

The earliest music promoters were in the publicity business at the beginning of the last century, primarily helping to sell sheet music that was heard on recording playback devices or at public performances. The title of the job during these times was that of "music pusher." Those who worked in the publicity profession in the early 1900s relied primarily on newspapers and magazines to promote the sale of music.

In 1922, the US federal government authorized the licensing of several hundred commercial radio stations, and those in the music business found their companies struggling as a result. People stopped buying as much music because radio was now providing it, and newspapers and magazines were no longer the only way the public got its news. Radio became the entertainer and the informer. But publicists found themselves with a new medium and a new way to promote, and quickly adapted to it, much in the way they did in 1948 with the advent of television as a news and entertainment form.

Today, the label publicist works with web, print, and television journalists for features. They also coordinate live appearances. Radio promotion handles airplay on AM, FM, and satellite radio, video promotion works with TV stations, cable channels, and websites that use videos as part of their programming and entertainment content and new media departments coordinate their work with publicity and promotion.

THE LABEL PUBLICITY DEPARTMENT

The work of a label publicity department (sometimes called the *media relations department*) is very much like a sales department. Rather than selling "things," they are selling ideas and an image. In particular, they are "pitching," or selling story ideas to the media, attempting to stimulate an interest by the correspondent in writing or producing a story about the artist's recording.

Staffing the Publicity Department

Large labels often employ a director or manager of publicity. The full responsibility of media relations and publicity rests with the director. They are accountable to the media outlets they serve, and they are also accountable to the various departments at the label such as A&R, marketing, and sales. Sometimes that accountability stretches to the artist and the artist management team.

Structurally, some managers of publicity report directly to the president/chief operating officer of the label, and they are often called *director* or *vice president of publicity*, or some similar title. Some labels align the publicity department's accountability directly to the marketing department, while others make publicity a component of the creative services department. (Creative services typically handles design and graphics work for albums, web retailers, and point-of-purchase [POP] material, as well as general imaging for the recorded music project.) Those who are hired as independent publicity companies for label projects typically report to the vice president of marketing.

Staff publicists work for the director and handle the day-to-day planning, coordination, and execution of the work of the department. Labels that have a large roster can stretch the time and energy of staff publicists, and sometimes the limitations of staff time require the employment of independent publicists. It is not uncommon to find that staff people are handling ongoing publicity efforts for several active artists at the same time.

Large label publicity departments sometimes hire independent or freelance publicists because these publicists have relationships with key media gate-keepers. They are hired because they have important contacts that the label does not have, and they can be effective in reaching these media outlets on behalf of the artists of the label.

Independent and smaller record labels sometimes handle publicity in-house, but they often hire a publicity firm on a project-by-project basis, or pay a retainer fee in order to have access to their services as they are needed. The cost of paying an employee to handle publicity at a small label becomes a financial burden at times when business is slow. During down cycles in the business of the major labels, entire publicity departments might be dismissed and independent firms hired to handle the work for them.

TOOLS OF THE LABEL PUBLICITY DEPARTMENT

The Database

The publicity department creates and distributes communications on a regular basis, so the maintenance of a quality, up-to-date contact list is critical to the success of that communication effort. Some publicity departments maintain their own contact lists; some use their own lists, plus lists through a subscription service; others rely entirely on subscription database providers.

An example of a subscription, or "pay" service, for media database management is Bacon's MediaSource. Bacon's updates its online database daily with full contact information on media outlets and subjects on which they report. The value of maintaining a quality database by the publicity department is that the information enables them to accurately target the appropriate media outlet, writer, or producer. Services like Bacon's can literally keep a publicity department on target. Most labels and their independent publicist partners maintain several lists within their databases to assure they are not sending news releases to people who are not interested in the subject matter. These lists are used to send out press releases, press kits, promotional copies of the CD for reviews, and complimentary press passes to live performances.

The most effective way to reach media outlets is through email, because "it's inexpensive, efficient, and a great way to get information out very quickly" (Stark, Phyllis, personal interview). Though some outlets still prefer faxes, the challenge is to program the machine for delivery at a time that is convenient so that home-based journalists are not awakened in the middle of the night with the alert on a fax machine. A few media outlets still prefer regular mail, but the immediacy of the information is lost. An effective publicist learns the preferred form of communication for each media contact. Bacon's MediaSource provides some of that information, but it is always preferable to check with the journalist.

Internet distribution of press information from a label requires the latest software that will be friendly to spam filters at companies that are serviced with news releases. A spam filter is a computer program used by companies to prevent the sending of unsolicited email to company employees. The most reliable way to assure news releases and other mass-distributed information are received by a media contact is to ask about any spam filters, and ways to bypass them.

The Press Release

The press release is one of the continuing tools of professionals in the publicity profession. It is used to publicize news and events, and is a pared-down news story. Following are five examples of when a press release should be used:

1. To announce the release of an album
2. To announce a concert or tour
3. To publicize an event involving the artist or label
4. To announce the nomination or winning of an award or contest
5. To offer other newsworthy items that would be appealing to the media.

The press release should be written with the important information at the beginning. Today's busy journalists don't have time to dig through a press release to determine what it is about. They want to quickly scan the document to determine whether this is something that will appeal to their target audience.

The Anatomy of a Press Release

The press release needs to have a slug line (headline) that is short, attention-grabbing, and precise. The purpose or topic should be presented in the slug line. The release should be dated with contact information including phone and fax numbers, address, and email. The body of text should be double spaced.

The lead paragraph should answer the five W's and the H (who, what, where, when, why, and how). Begin with the most important information; no unnecessary information should be included in the lead paragraph (Knab, C., 2003a). In the body, information should be written in the inverse pyramid form: in descending order of importance.

The Bio

The artist biography provides a window into the artist's persona, and sets the artist apart from others. Before writing the bio, an examination should be done on the artist's background, accomplishments, goals and interests to find interesting and unique features that will set the artist apart from others.

Knab, C. (2003b)

Keep in mind the target readership of the bio. Some busy journalists may use portions of the bio in a news story. The bio should be succinct and interesting to read (Hyatt, 2004). Create an introduction that clearly defines the artist and the genre or style of music.

The Press Kit

Another of the primary tools of the department is the press kit they develop on behalf of the artist and the recorded music project. Typical components of physical and digital press kits include:

- A press release announcing the release of the CD
- The artist's bio, which is often created by someone hired by the label for this specific purpose
- The CD that contains the single or the album
- High-resolution color jpegs of the artist on disc
- A "cut-by-cut," which is a document attached to the press release listing of the tracks included on the CD with the artist's personal comments about each track
- A discography
- Sometimes it also includes an electronic press kit (EPK), which is a copy of the artist's music video and also shows them discussing the music and the project. This helps journalists get a sense of what the artist looks like and how they relate other than through their music
- Clippings (tear sheets): articles printed or video stories created by other media.

Journalists say that the most useful parts of the press kit, sometimes called a *pitch kit*, are the CD, the cut-by-cut, and the photos.

A cover letter or email is included with the press kit attached, and should be personal to the journalist. The purpose of the communication must be very clear in the first paragraph, and it should be specific, not general. From the journalist's perspective, answer the questions: "What do you want?" and "Why is this important to me?" Otherwise, the label publicist risks having the journalist set the kit aside to, perhaps later, figure out what it is all about. And "later" may never happen.

©2009 Paul Allen

Photos and Video

A good photograph can generate a lot of publicity. It can be the most striking and effective part of a press kit. A good quality photo has a much better chance of being run in print media and is worth the extra effort and expense. An experienced professional photographer can bring out the true personality of the artist in a photograph. Sometimes, location is used to help portray the artist's identity, but studio shots are easier to control (Knab, C., 2001). Publicity shots are not the same as a publicity photo. A publicity shot is one taken backstage with other celebrities, or at events. The publicity photo is the official photographic representation of the artist. Publicity photos should be periodically updated to keep current with styles and image. But, once a photo is released to the public, it is fair game for making a reappearance at any time in the artist's career, even if they have moved on and revamped their image.

The music video also becomes an important part of the continuing effort of the publicity department to promote the album as consecutive singles are released. Though many labels have video promotion and "new media" departments, the artist's music video is also a valuable tool used by publicists in securing live and taped appearances in television programs.

Working with the Artist's Image

Any aspect of the entertainment business relies on the created perception of the artist or event. After all, it is show business. It is important to the label to know the perception of the recording artist in the minds of music buyers. One of the most important contributions a label publicity department can make towards defining the public's perception of the artist is carefully helping the artist develop their image.

There are several key items that contribute to the image of the artist. Among those are:

- The name chosen by the artist
- Physical appearance of the artist
- Their recording style and sound
- Choices of material and songwriting style
- Their style of dress
- The physical appearance of others who share the stage
- The kind of interviews done on radio and TV
- Appearance and behavior when not on stage (Frascogna, X., and Hetherington, L., 2004).

Some labels will hire hair stylists and clothing and costume consultants, some will pay for dental work and some are rumored to pay for cosmetic surgery in order to polish images for their expanded public career (Levy, S., 2004).

Hiring a media consultant is a judgment call for the label, and they may or may not feel that an artist should receive training. What a media consultant does is train artists to handle themselves in public interviews. Working with the artist, the consultant prepares the artist for interviews by taking the unfamiliar and making it familiar to them, teaching them to know what to expect, and giving them the basic tools to conduct themselves well in a media interview. The debate comes from critics who say media consultants go too far by preparing artists with suggested answers, and thereby create cookie-cutter interviewees with nothing new or interesting to say.

Publicists and others at the label should take care not to compromise the unique qualities an artist has by developing an image that isn't consistent with who the artist is. The values the artist has should be apparent in their music and public image in order to tighten the connection with record consumers who share those values. Ideally, the label and the artist manager will work with the artist to help them define who they are personally and creatively, and then coach them to an image that is commercial but not artificial.

For established artists, care must be taken not to radically change their image. Fans are quick to pick-up on efforts to radically re-tool an image, and the result can ultimately turn fans away from an artist. However, there are arguable exceptions to this rule for more established artists. For example, Madonna has modified her image across time, sometimes from album to album, albeit not drastically, and continued her success in 2008 with another number one album.

Evaluation of Publicity Campaigns

Top management and artists themselves will seek feedback about the effectiveness of a media campaign on behalf of an album project. Clipping services are available that will search publications and television shows for copies of articles or video clips that will demonstrate the news item connected with the media. AristoMedia, an independent publicity company in Nashville, often uses Google as an efficient way to check the impact of its news releases a few days following dissemination. It is good for artist relations to present the artist and manager with a box of news clippings and files from the web to indicate the successful efforts of the publicity department in securing media coverage.

TARGET MARKETING THROUGH THE PUBLICITY PLAN

Label publicity, as in any part of the marketing process, must keep its focus on the target market for the recording, tour dates, and availability of ancillary merchandise. The starting point is to get a clear understanding of what that target market is and how publicity fits within the project's marketing plan, then to develop a plan to reach the market through publicity. The plan created by a publicity department is coordinated with other departments within the label. For example, A&R will define the artist, the music, and the genre. Marketing will provide the overall plan that will include sales objectives, the target market, the marketing communications plan, as well as an estimated release date for the single and the album. The new media department will coordinate web promotion. Radio promotion will provide its timeline to work the record to radio, and let publicity know if the music will be promoted to multiple formats. That helps the publicist know where to place their efforts. Sales will give its timetable so that publicity efforts will be timed to maximize sales at retail.

Table 9.1 Media and Target Examples	
Media	**Targets and Examples**
Daily newspapers	General readers from a national, regional, local base
Lifestyle, entertainment magazines	Broad national readership with a certain editorial focus *People, Entertainment Weekly, Cosmopolitan*
News weeklies	Young, urban, sophisticated, culturally aware readers *Village Voice, SF Weekly, Boston Phoenix*
Music and pop culture magazines	Young, affluent adults with strong interest in … music *Rolling Stone, Spin, Vibe*
Genre-based music magazines	Customers of chain record stores *Country Weekly, The Source, Flipside*
Promotional magazines	Customers of chain record stores *Pulse, Request*
Magazines for music hobbyists and pros	Players of instruments, recording engineers *Guitar World, Mix, Musician*
Magazines for record collectors	Collectors and afficionados of oldies and rare discs *Goldmine, DISCoveries*
Trade publications	Music industry professionals *Billboard, Radio & Records, Variety*
Fanzines	Pre-teen and teenage fans of new artists *Tiger Beat*

Source: *This Business of Music Marketing and Promotion*, Lathrop and Pettigrew (1999).

Reaching the target audience with press releases requires an understanding of the actual audience of the media. For example, Table 9.1 outlines the traditional media type and its intended audience.

Trade publications, like those noted above, offer the benefit of setting up a recorded music project to garner the interest of radio and video channels. New artists especially require a major push by publicity to create that invaluable buzz among the electronic gatekeepers of radio, television, and the Internet. The trades must be worked before the release of the album, and radio should be worked prior to the release of the first single. Press releases can be sent to traditional and electronic trade publications with the objective of getting an article or at least a blurb about the upcoming release.

The savvy publicist researches the magazines, newspapers, fan magazines, and online entertainment sources before submitting materials for publication. The types of stories, reviews, and features that each publication uses should be noted. Then, only those that match the target market and that regularly feature the types of stories being pitched should be approached. Nothing aggravates a writer or editor more than someone pitching inappropriate material to their publication. It is generally easier to get placement in a niche magazine than a more general publication. Genre-based music magazines and websites are more receptive to publishing suitable material, and there is less competition than one would find with a general interest publication.

The Publicity Plan

Peter Spellman says, "the first ingredient for a successful publicity plan is a clear idea of your market audience: who they are, what they read and listen to, and where they go. Each style of music is a subcultural world ... your job is to understand this world" (Spellman, P., 2000a). The publicity plan is designed to coordinate all aspects of getting non-paid press coverage and is timed to maximize artist exposure and record sales. The plan is usually put into play several weeks before the release of an album. Spellman also says that "your publicity objectives can only be realized through successive 'waves' of media exposure" (Spellman, P., 1996b) because of the short attention span of the public. Each wave needs a promotional angle: an album release, announcement of tour dates, or anything newsworthy (Table 9.2).

The publicity plan starts with goals: awareness, motivation, sales, positioning, and so on, and priorities are established (Yale, D.R., 2001). The marketplace is then researched, and media vehicles targeted. The materials are developed and the pitching begins. Lead time is the amount of time in advance of the publication that a journalist or editor needs to prepare materials for inclusion in their publication. A schedule is created to ensure that materials are provided in time to make publication deadlines. Long-lead publications are particularly problematic for the publicist as they need to have materials prepared months in advance of the release date, and sometimes those materials are not yet available. If an artist suddenly breaks in the marketplace, it is too late to secure a last-minute cover photo on most monthly publications.

Pitch letters are then sent out to media requesting publication or other media exposure. The pitch letter is a carefully thought-out and crafted document specifically

Table 9.2

- Set publicity goals
- Identify target market
- Identify target media
- Create materials
- Set up timetable w/deadlines
- Pitch to media
- Provide materials to media
- Evaluate

Source: The Publicity Handbooks, Yale (2001).

designed to grab the interest of a busy, often distracted journalist or TV producer. It is never mass-mailed, but is specifically tailored to each media outlet being contacted (D'Vari, M., 2003). The pitch letter should begin with a few words presenting the publicist's request, and then quickly pointing out why the media vehicle being contacted should be interested in the artist or press material. Prep sheets are also developed and sent to radio so that DJs can discuss the artist as they spin the record. As the publicity plan unfolds, it is necessary to evaluate the efforts through clipping services and search engines (see section on evaluation of publicity campaigns).

Budgets for Money and Time

A budget for the publicity campaign is developed based on the objectives of the project and the expectations of the label for the part publicity will play in stimulating interest in the artist's music. If this is the first album for the artist, the develop-

Table 9.3 Example Publicity Timeline for a Major Label	
Time Frame	**Publicity Task**
Upon signing the artist	Schedule meetings with artist Press release announcing signing
During the recording	In-studio photos Invite key media people to studio
Also during this time period	Schedule media training if needed Select media photos Determine media message
When masters are ready	Hire bio writer Create advance copies for reviews Create visual promo items
When advance music is ready—Ideally four months out	Send advances to long-lead publications Send advances w/bio and photo to VIPs—magazines, TV bookers, and syndicators Begin pitch calls to secure month-of-release reviews Start servicing newsworthy bits on the artist on a weekly basis to all media
Advance Music—one month out	Send advances/press kits to key newspaper and TV outlets Begin pitch calls to guarantee week-of-release reviews
One week out	Service final packaged CDs to all media outlets Continue follow-up calls and creative pitching
After release	Continue securing coverage and providing materials to all media outlets

Source: Amy Willis, Media Coordinator, Sony Music Nashville.

©2009 Paul Allen

Table 9.4

Day	Date	Event
Fri	November 2	Advance promo copies at *The Rocket, UW Daily, Pandemonium, Seattle Times*, and the *Seattle Weekly*
Mon	November 3	Arrange interviews on KCMU, Seattle and KUGS Bellingham
Fri	December 13	Deadline for completing database of print/broadcast mailing list
Fri	January 10	Single sent to College radio
Fri	January 17	*Calling It a Day* CD release day
Mon	January 20	Album mention in *UW Daily/Seattle Times*
Tues	January 21	Album mention in *Seattle Weekly*
Thurs	January 24	Tour begins in Seattle
Thurs	January 24	Album mention in *CMJ/Hits*
Tues	January 28	Album mention in the *Rocket and Pandemonium*
Tues	January 27	Album mention in Spokane and *Tri-City Daily* papers
Mon	January 30	Album mention in *Cake and Fizz*
Fri	January 3	Album mention in *Flipside* and *Village Noize*
Wed	February 15	Album mention in Spin
Wed	February 15	Album mention in *Virtually Alternative*
Fri	February 17	Album mention in *Next*
Fri	January 12	Album mention in *Magnet*
Mon	February 27	Interview on KUGS Bellingham
Mon	February 27	Interview on KCMU Seattle
Tues	January 14	Feature story in *The Rocket*

Source: Christopher Knab, www.4frontmusic.com

ment of new support materials may be necessary, such as current photos and a bio. If it is an established artist, budgets could be considerably higher, in part because of the expectations of the artist to receive priority attention from the director of publicity (Tables 9.3 and 9.4).

Publicity costs include the expense of developing and reproducing materials such as press kits, photos, bios, and so forth, communication costs (postage and telephone bills, contact lists), and staffing costs. The minimum cost for an indie label would run about $8,000, with $3,000 of that for developing press kits and $2,800 for postage. Adding an outside consultant to the project would add another $1,500 or more per month. For major label projects, an outside publicist can be hired for 6 months to provide full support to a single, and album, and tour publicity for $25,000, which includes out-of-pocket costs such as postage, press kits, and other related expenses.

An equally important part of the plan is to budget adequate time to support the album based on when it will be released during the annual business cycle of the label. If the in-house staffing is adequate—given the timing of the project—the plan can be created. If, however, the publicity department is overloaded, the director may consider hiring an independent company to handle publicity for the project. This seemingly removes the burden from the director, but it adds oversight duties since the director must be sure the outside company is working the plan according to expectations. The ultimate success (and failure) is still the responsibility of the director of publicity.

TELEVISION APPEARANCES

News Shows

Major entertainment television news shows, including syndicated news shows on major network affiliates, are most often interested in major acts. They see their viewers as people who want to know about the latest information on their favorite recording artists. Stars that are easily recognizable are those most often sought for their entertainment news stories. This creates a genuine challenge for the record label that is trying to publicize a new artist with consumers. A new act with a new single or album must have an interesting connection with consumers that goes beyond the music in order to compete with the superstars who will always get airtime. There are many more new artists looking for publicity than there are established artists, and it forces the best label publicists to be as creative as they can be on behalf of the few new acts.

For television interviews with new artists, a sit-down interview with the artist often is not enough. Television shows look for that added dimension to a new artist that makes them interesting to the viewers, and they often look for the non-traditional setting in which to present the story. Though at times it is overdone, connecting an artist with their charity work becomes an interesting angle for television.

The challenge to the label publicity department is to find those key personal differences, or "back stories," that make their recording artists interesting beyond their music. Label publicists are sometimes criticized for citing regional radio airplay, chart position, or label financial support as the only positives that make their newest artists stand out. Those in the media say they look for that something special, different, and newsworthy that gives an angle for them to write about. In that light, it puts the responsibility on the label publicist to find several different angles to offer to different media outlets to generate the interest needed to get a story placed. Writers for major media want their own angle on an artist when possible because it demonstrates to media management that an independent standout story has been developed, making them different from their competition. Sometimes, though, the story angle about an artist is different enough that it stands on its own and most media will see the value it has for their audiences. Entertainment writers and producers are often self-described storytellers, and delivering that unique story to them is a continuing challenge to the successful label publicist (Pettigrew, J., Jr., 1997).

Talk–Entertainment Shows

Label publicists are often the facilitator of an artist's appearance on popular talk and entertainment shows. Among the most popular shows include those with hosts David Letterman, Conan O'Brien, Oprah, and Regis Philbin. Additional television shows that publicists seek as targets for the label's artists include the early morning shows like "Today" and "Saturday Night Live." Great performances on shows like these are expected of the artist, but Ashley Simpson's ill-fated 2004 appearance on "Saturday Night Live" will probably follow her forever. (Simpson had an embarrassing performance when tracks for the wrong song were cued up and played, revealing pre-recorded vocal tracks.)

Bookings to programs like these are usually handled by the publicist and are based upon their relationships with each show's talent bookers. It is not uncommon for a publicist to precede a pitch for an artist to appear on one of these shows by sending a big fruit basket. However, the success of placing the label's new artist on one of these shows is also based on the ability of the publicist to build a compelling story for the artist that will interest the booker. Often the publicist will offer another major artist for a later appearance in exchange for accepting the new artist now.

Award Shows

The value of having an artist perform on an award show is obvious—it sells records. These slots are coveted by all the record labels, and lobbying efforts may pay off in a big way.

COMPARING PUBLICITY AND RECORD PROMOTION

The savvy labels recognize how important publicity is to the mix—it's almost as important as record promotion.

Phyllis Stark, Former Nashville Editor,
Billboard Magazine

Record promotion is something that goes hand in hand with publicity and will be synchronised to make the maximum impact in the marketplace. In Table 9.5 we look at the relationship that publicity has with its counterpart in the effort of promoting an album.

The connection between recorded music and radio is essential to the success of large record labels that compete in the national and international music mass markets. The charts both here and in the UK are a driving force in the exposure for artists and ultimately sales.

In the US the two major magazines for the music industry are *Billboard* and *Radio & Records*. Both provide comprehensive weekly views of the recording industry, the

Table 9.5   Comparison of Promotion and Publicity Departments

Record Promotion	Publicity Department
Develops and maintains relationships with key radio programmers (gatekeepers).	Develops and maintains relationships with key writers, news program producers, and key talent bookers for network and cable channel TV shows.
Tells radio programmers that a new single or album is about to be released and to prepare for "add" date; sends promo singles and albums.	Prepares and sends a press kit to journalists announcing the new single or album project.
Schedules the new artist for tours of key radio stations for interviews and meet 'n greets with station personnel.	Schedules the artist and sometimes the album producer for interviews with both the trade press and consumer press.
Employs independent radio promotion people who have key relationships with important radio programmers.	Employs independent or freelance publicists who have key relationships with important media outlets.
Effectiveness of their work is measured by the number of "adds" they receive on the airplay charts of major trades.	Effectiveness is measured by the number of "gets" they receive, meaning the number of articles placed, number of TV news shows in which stories run, the number of talk/entertainment shows on which the artist performs (Phyllis Stark).
Gets local radio publicity and airplay for new artists based on the promise of an established artist making a local appearance sometime in the future.	Gets new artists booked on major talk/entertainment shows based on the promise of making an established artist available to the show sometime in the future. Supports local press during touring.

music business, and commercial radio. These publications publish the charts and the overall success of artists in the marketplace in their respective territories.

GETTING A RECORDING ON THE RADIO

The consolidation of radio in the US has concentrated some of the music programming decisions into the hands of a few programmers who provide consulting and guidance from the corporate level to programmers at their local stations. In many cases, local programmers have the ability to add songs to their playlists based upon the preferences of the local audiences. Here is how songs are typically added to station's playlists for those stations that program new music:

- A record label promoter or an independent promoter hired by the label calls the station music director (MD), or program director (PD) announcing an upcoming release. Radio music directors have "call times." These are designated times of the week that they will take calls from record promoters. The call times vary by station and are subject to change. For example, an MD may have call times of Tuesdays and Thursdays, 2:00–4:00 p.m.

- Leading up to the add date, meaning the day the label is asking that the record be added to the station's playlist, the promoter will call again touting the positives of the recording and ask that the recording be added.
- The music director or program director will consider the selling points made by the promoter, review the trade magazines for performance of the recording in other cities, consider current research on the local audience and its preferences, look at any guidance provided by their corporate programmers/consultants, and then decide whether to add the song.
- The PD will look for reaction or response to adding the song. The "buzz factor" for a song will be apparent in the call-out research and call-in requests, as well as through local and national sales figures – especially digital download sales.

An important component of promoting a recording to radio is the effectiveness of the record company promotion department or the independent promoters hired to get radio airplay. This would appear to be a simple process but the competition for space on playlists is fierce. Thousands of recordings are sent to radio stations every year, and the rejection rate is high because of the limited number of songs a station can program for its audience. Some of the recordings are rejected from being included on playlists because they are inferior in production quality, some are inappropriate for the station's format, and many lose their label support if they fail quickly to become commercial favorites with the radio audience.

However, the marketer's litmus test for the viability of a recording is to honestly compare it with other songs on the charts of trade magazines. If it is not at least as good as those listed on the current trade magazine charts, then it doesn't have any chance at all, even if it has a competitive marketing budget.

INDEPENDENT PROMOTION

Independent record promoters are contractors who work at the direction of the label's vice president of promotion to augment album promotional efforts at terrestrial radio. Since promoting songs for airplay relies on well-developed relationships, "indies" may have developed stronger relationships with some key stations than the label has, and the record company is willing to pay indies (half of which is often recoupable from the artist) for the value of those relationships.

A scandal in the 1980s was about the practice of labels hiring independent promoters, or "indies" to attain airplay at radio. Fred Dannen, in his book, *Hit Men*, found that CBS Records was paying $8–10 million per year to indies to secure airplay for their acts. By the mid-1980s he says that amount was $60–80 million for all labels combined. Then in 2005 and 2006 the New York attorney general successfully investigated and settled with all of the major record labels for violations of the payola law, and indie promotion has once again become dormant.

Label Record Promotion and Independent Promoters

Most large labels have a promotion department whose sole purpose is to achieve the highest airplay chart position possible. While most consumers assume a number one song is the biggest seller at retail, the number one song on most Billboard and R&R charts actually is the song that has the most airplay on radio. The connection between traditional radio airplay and sales is well-documented, so a high chart position is critical to the success of a recording and becomes the heart of the work of a record promotion department.

Labels often have a senior vice president of promotion who usually reports directly to the label head. The senior vice president of promotion at a pop music label typically has several vice presidents of various music types based on radio formats. These vice presidents then have regional promotion people who are viewed by the label as field representatives of the promotion department. They are liaisons to key radio stations in their region. The vice president and the so-called regionals are the front line for the label attaining airplay. It is their responsibility to create and nurture relationships with programmers for the purpose of convincing them to add the company's recordings to their playlists. At a country label, the senior vice president of promotion typically has a national director of promotion and several regionals.

ACKNOWLEDGMENT

A special thanks to Tom Baldrica, Bill Mayne, Jeff Walker, Phyllis Stark, Ed Benson, the folks at Starpolish.com, and the CMA for their assistance in providing added information and insight into this chapter.

GLOSSARY OF PUBLICITY TERMS

Bio Short for biography. The brief description of an artist's life and/or music history that appears in a press kit.

Clippings Stories cut from newspapers or magazines.

Creative services department A work unit at a record label that handles design, graphics, and imaging for a recorded music project.

Cut-by-cut This is a listing of comments made by an artist about each of the songs chosen to be included on an album project.

Discography A bibliography of music recordings.

Independent publicist Someone or a company that performs the work of a label publicist on a contract or retainer basis.

Lead time Elapsed time between acquisition of a manuscript by an editor and its publication.

Media consultant Trains artists to handle themselves in public interviews with the media.

Music pusher A term used in the early 1900s for someone who was promoting music.

Press kit An assemblage of information used to provide background information on an artist.

Press release A formal printed announcement by a company about its activities that is written in the form of a news article and given to the media to generate or encourage publicity.

Slug A short phrase or word that identifies an article as it goes through the production process; usually placed at the top corner of submitted copy.

Talent bookers These are people who work for producers of television shows whose job it is to seek appropriate artists to perform on the program.

Tear sheets A page of a publication featuring a particular advertisement or feature, and sent to the advertiser or PR firm for substantiation purposes.

REFERENCES

This edited chapter is from Hutchinson, T. et al. (2006). *Record Label Marketing*, Focal Press, Burlington, US.

D'Vari, M. (2003). www.publishingcentral.com/articles/20030301-17-6b33.html. *How to Create a Pitch Letter*.

Frascogna, X. & Hetherington, L. (2004). *This Business of Artist Management*. Billboard Books, New York.

Hyatt, A. (2004). www.arielpublicity.comhttp://arielpublicity.com. *How to be Your Own Publicist*.

Knab, C. (2001). www.musicbizacademy.com/knab/articles/. *Promo Kit Photos*.

Knab, C. (2003a). www.musicbizacademy,com/knab/articles/pressrelease.htm. *How to Write a Music-Related Press Release* (November, 2003).

Knab, C. (2003b). www.musicbizacademy.com/knab/articles.

Lathrop, T. & Pettigrew, J., Jr. (1999). *This Business of Music Marketing and Promotion*. BPI Publications, New York.

Levy, S. (2004). CMA's music business 101, unpublished.

Pettigrew, J., Jr. (1997). *The Billboard Guide to Music Publicity*. Watson-Guptill Publications, New York.

Spellman, P. (Spring 1996b). www.musicianassist.com/archive/newsletter/ MBSOLUT/files/mbiz-3.htm. *Creative Marketing: Making Media Waves: Creating a Scheduled Publicity Plan*. Music Business Insights (Issue I:3).

Spellman, P. (2000a). www.harmony-central.com./Bands/Articles/Self-Promoting_ Musician/chapter-14-1.html. *Media Power: Creating a Music Publicity Plan That Works, Part 1* (March 24, 2000).

Yale, D.R. (2001). *The Publicity Handbook*. NTC Business Books, Chicago.

Chapter 10
Music Distribution and Retail

Amy Macy

In This Chapter

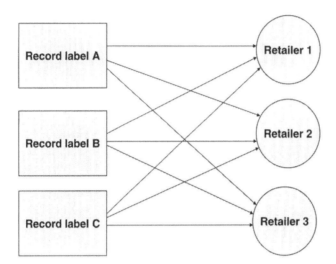

FIGURE 10.1
Direct contact
concept.

INTRODUCTION

Music distributors are a vital conduit in getting physical and digital music products from record labels' creative hands into the brick-and-mortar and virtual retail environment. Recognize that as the marketplace shifts to a digital environment, distribution companies are re-evaluating their value in the food chain and continually developing sales models that reflect direct sales opportunities to music consumers. By the end of 2007, nearly 1 out of every 4 albums purchased were at "non-traditional" outlets which include digital services, Internet retailers, mail order, at "non-traditional stores" or at a concert. Whatever the sales channel, distribution companies think of themselves as extensions of the record labels that they represent.

Prior to the current business model, most individual record labels hired their own sales and distribution teams, with sales representatives calling on individual stores to sell music. It took many reps to cover retail, as Figure 10.1 shows. This model shows nine points of contact, where each label meets with each retailer. As retailers became chains, and as economics of the business evolved, record labels combined sales and distribution forces to take advantage of economies of scale, which eventually evolved into the current business model.

Figure 10.2 includes the distribution function and shows the points of contact reduced to six. In today's business model, record label sales executives communicate with distribution as their primary conduit to the marketplace, but labels also have ongoing relationships with retailers. Depending on the importance of the retailer, the label rep will often visit the retailer with their distribution partner so that significant releases and marketing plans can be communicated directly from the label to the retailer. And as deals are struck, both orders and marketing plans can then be implemented by the distributor.

©2009 Amy Macy

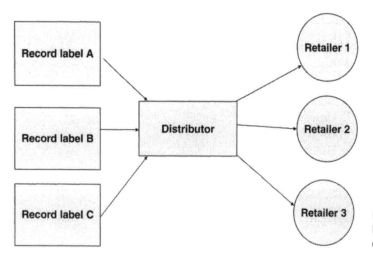

FIGURE 10.2
Distribution centered concept.

THE BIG 4 AND MORE

Within the last 10 years the Big 4 have consolidated from the Big 6, reducing their numbers of employees to reflect the ever-decreasing size of the music sales pie. But the mergers began prior to the explosion of file sharing with the combination of Universal with Polygram in 1999, the same year as the emergence of Napster™. This merger created UMVD who has maintained their market share position at #1 since inception.

More recent is the combination of Sony and BMG in 2004, who consolidated their conglomerates as well as their distribution workforces but have maintained much of the integrity of their imprints. They, too, have gained market share positioning by garnering the #2 position simply by merging. But as of August 2008, Sony is in negotiations of purchasing the BMG holdings of the partnership.

WEA and EMI maintain separate distribution functions as part of Warner Music Group and EMI Group respectively. Although there has been speculation that the two would consider joining forces to combat the larger entities, no deal has been struck and EMI was most recently purchased in 2007 by the private equity firm Terra Firma, causing many to rethink the merger/acquisition possibility.

As noted in the market share data, independent labels continue to be a driving force within the music business and in 2007, independents broke the 20% market share threshold. With the burgeoning digital storefronts, any label can now have an instant "sales" point in which to connect with customers directly. But to fulfill physical product, independent labels need to reach the brick-and-mortar outlets using traditional methods. Most of the major distributors have created an independent arm of their distribution function within their family. By contracting this function of distribution to independent labels, these "independent distributors" can assist the independent label in placing their music in the mainstream marketplace. But there are many true independent distributors that are *not* tied to the Big 4 that function similarly by placing physical product in stores.

UMVD—Universal Music and Video Distribution

Sample labels that UMVD distributes:

Interscope
Geffen
Island/Def Jam
Universal
UMG Nashville
Hollywood
Disney/Buena Vista

Sony BMG

Sony's RED Distribution represents many independent labels.

Sample labels that Sony BMG distributes:

Columbia
Epic
Arista
J Records
RCA
Jive
LaFace
Razor & Tie
WindUp
RCA Label Group Nashville

WEA

WEA's ADA Distribution represents many independent labels.

Sample labels that WEA distributes:

Warner Brothers
Atlantic Records
Bad Boy
Roadrunner
WSM/Rhino
V2 Records

EMD

EMD's Caroline Distribution represents many independent labels.

Sample labels that EMD distributes:

Blue Note
Capitol Nashville
Capitol Records
Virgin
EMI Latin

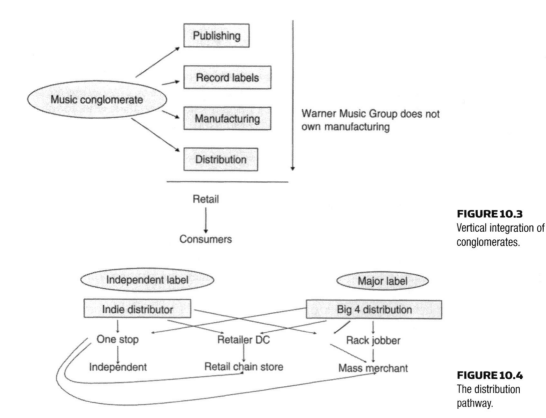

FIGURE 10.3
Vertical integration of conglomerates.

FIGURE 10.4
The distribution pathway.

VERTICAL INTEGRATION

Three out of the Big 4 conglomerates (major labels) share profit centers that are vertically integrated, creating efficiencies in producing product for the marketplace. To take full advantage of being vertically integrated, labels looking for songs would tap their "owned" publishing company. (Each of the major labels has a publishing company. If they only recorded songs that were published by their sister company, more of the money would stay in-house.) Once recorded, the records would be manufactured at the "owned" plant. In turn, the pressed CDs would then be sold and distributed into retail—with the money being "paid" for each of the functions staying within the family of the music conglomerate (Figures 10.3 and 10.4).

In addition to the Big 4 companies, again, there are many independent music distributors that are contracted by independent labels to do the same job. Ideally, the distribution function is not only to place music into retailers, but also to assist in the sell-through of the product throughout its life cycle, and independent distributors have developed a niche in marketing unique and diverse products.

MUSIC SUPPLY TO RETAILERS

Once in the distributor's hands, music is then marketed and sold into retail. Varying retailers acquire their music from different sources. Most mass merchants are serviced by *rack jobbers*, who maintain the store's music department including inventory management, as well as marketing of music to consumers. Retail chain stores are usually their own buying entities, with company-managed purchasing offices and distribution centers (DCs). Many independent music stores are not large enough to open an account directly with the many distributors, but instead work from a *one-stop's* inventory as if it were their own. (One-stops are wholesalers who carry releases by a variety of labels for smaller retailers who, for one reason or another, do not deal directly with the major distributors.) Know that retail chain stores and mass merchants will on occasion use one-stops to do "fill-in" business, which is when a store runs out of a specific title and the one-stop supplies that inventory. (Retail is discussed in detail later in this chapter.)

ROLE OF DISTRIBUTION

Most distribution companies have three primary roles: the sale of the music, the physical distribution of the music, and the marketing of the music. Reflecting the marketplace, most distribution companies have similar business structures to that of their customers—the retailers. Oftentimes, the national staff is separated into two divisions: sales and marketing. The sales division is responsible for assisting labels in setting sales goals, determining and setting deal information, and soliciting and taking orders of the product from retail. Additionally, the sales administration department should provide and analyze sales data and trends, and readily share this information with the labels that they distribute.

The marketing division assists labels in the implementation of artists' marketing plans along with adding synergistic components that will enhance sales. For instance, the marketing plan for a holiday release may include a contest at the store level. Distribution marketing personnel would be charged with implementing this sort of activity. But the distribution company may be selling holiday releases from other labels that they represent. The distribution company may create a holiday product display that would feature all the records that fit the theme, adding to the exposure of the individual title.

The physical warehousing of a music product is a huge job. The major conglomerates have very sophisticated inventory management systems where music and its related products are stored. Once a retailer has placed an order, it is the distributor's job to pick, pack, and ship this product to its designated location. These sophisticated systems are automated so that manual picking of product is reduced, and that accuracy of the order placed is enhanced. Shipping is usually managed through third-party transportation companies.

As retailers manage their inventories, they can return music product for a credit. This process is tedious, not only making sure that the retailer receives accurate credit for product returned, but the music itself has to be retrofitted by removing stickers and price tags of the retailer, re-shrink wrapping, and then returning into inventory.

Conglomerate Distribution Company Structure as It Relates to National Retail Accounts

Although there are many variations and nuances to these structures as determined by each distribution company, the basic communication chart in Figure 10.5 still

©2009 Amy Macy

FIGURE 10.5
Conglomerate distribution company structure.

applies. At each level, the companies are communicating with each other. At the national level, very complex business transactions are being discussed, including terms of business as well as national sales and marketing strategies that would affect both entities company-wide. As mandated by the national staff, the regional/branch level is to formulate marketing strategies, either as extensions of label/artist plans or that of a distribution focus. At the local level, implementation of all these plans is the spotlight. But by design, these business structures are in place to create the best possible communication at every level, with an eye on maximizing sales.

NATIONAL STRUCTURE

To optimize communication along with service, distributors need to be close to retailers. Many of the major conglomerates have structured their companies nationally to accommodate the service element of their business. Most distributors have regional territories of management containing core offices and distribution centers (DCs). Each region contains satellite offices, getting one step closer to actual retail stores. As reflected in the structure chart in Figure 10.6, field personnel are on the front line, reporting to satellite offices, who then report to the regional core offices.

This U.S. map shows basic regions for a major distributor with satellite offices and distribution centers. Core regional offices are located in Los Angeles, Chicago, and New York. Regions are naturally eastern, central, and western, with regional offices including Atlanta, Detroit, Minneapolis, Dallas, and Seattle. The DCs are centrally located within each region. In this example, the DCs are located in Sparks, NV; Indianapolis, IN; and Duncan, SC. No destination is more than a 2-day drive from the DC, making product delivery timely.

Although the core regional offices are centrally located, these offices are primarily marketing teams, executing plans derived by both the labels that they represent and the distribution company. To be clear, over 75% of the music business is purchased and sold through 10 retailers. The locations of these buying offices are key sales sites, and designated distribution personnel are placed near these retailers so that daily, personal interaction can occur. These locations and retailers are shown in Table 10.1.

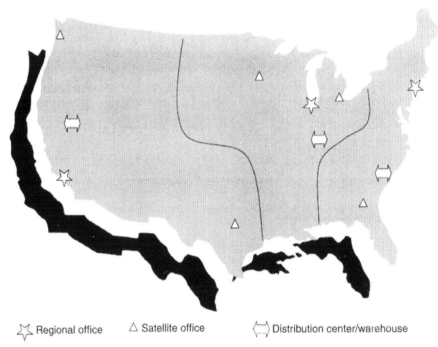

△ Regional office   △ Satellite office   ⊂⊃ Distribution center/warehouse

FIGURE 10.6
U.S. map of distributor locations.

Table 10.1	
Bentonville, AR	Wal-Mart
Minneapolis	Best Buy and Target
Miami	Alliance
Albany	Transworld

TIMELINE

The communication regarding a new release begins months prior to the street date. Although there are varying deadlines within each distribution company, the ideal timeline is pivotal on the actual street date of a specific release. For most releases, street dates occur on Tuesdays. Working backwards in time, to have product on the shelves by a specific Tuesday, product has to ship to retailer's DCs approximately 1 week prior to street date. To process the orders generated by retail, distributors need the orders 1 week prior to shipping. The sales process of specific titles occurs during a period called *solicitation* (Table 10.2). All titles streeting on a particular date are placed in a solicitation book, where details of the release are described. The solicitation page, also known as *Sales Book Copy*, usually includes the following information:

- Artist/title
- Street date

Table 10.2	Sales Timetable	
Prior to SD	**Activity**	**Example Dates**
8 weeks	Sales book copy due to distributor	September 27
6 weeks	Solicitation book mailed to retail buyers	October 11
4 weeks	Solicitation	October 25 to November 5
2 weeks plus	Orders due	November 5
1 week	Orders shipped wholesalers/retail chain	November 16 orders received
5 days	Orders shipped to one-stops	November 19 orders received
Street date		November 23

- File under category—where to place record in the store
- Information/history regarding the artist and release
- Marketing elements:
  - Single(s) and radio promotion plan
  - Video(s) and video promotion plan
  - Internet marketing
  - Publicity activities
  - Consumer advertising
  - Tour and promotional dates
- Available POP
- Bar code

This information is also available online on the business-to-business (B2B) sites established for the retail buyers.

ONE SHEET—THE SOLICITATION PAGE

On the website, *MusicDish Industry e-journal*, Christopher Knab of ForeFront Media and Music describes a one sheet as:

> A Distributor One Sheet is a marketing document created by a record label to summarize, in marketing terms, the credentials of an artist or band. The One Sheet also summarizes the promotion and marketing plans and sales tactics that the label has developed to sell the record. It includes interesting facts about an act's fan base and target audience. The label uses it to help convince a distributor to carry and promote a new release.

Knab (2001)

FIGURE 10.7
One sheet for
Buck 65. (*Source*: V2
Records.)

The one sheet typically includes the album logo and artwork, a description of the market, street date, contact info, track listings, accomplishments, and marketing points. The one sheet is designed to pitch to buyers at retail and distribution. The product bar code is also included to assist in buy-plugging the actual release into the inventory code system (Figure 10.7).

©2009 Amy Macy

DIGITAL DISTRIBUTION

According to SoundScan, year-end 2007 data reveals that 23% of music was purchased at "non-traditional" retailers, with the remaining 78% through brick-and-mortar stores.

The primary function for music distributors is now split between that of physical distribution of CDs into physical stores and assisting in the relationships between third-party licensing sites. In some cases, many of the major distributors have sites where consumers can buy tracks and albums directly from the conglomerate. These sites usually promote and sell music from their represented labels only, which makes it difficult for consumers to experience one-stop shopping.

In turn, within most conglomerate families resides a department that licenses music to third parties, which are legal downloading sites such as iTunes®, Rhapsody, and Napster™, plus the ever-popular ringtone and ringback services that are provided by mobile operators such as Sprint, Verizon, and LG. These licensing departments are critical to the evolution of distribution, positioning them as gatekeepers for the growing downloading environment. At these third-party sites, consumers can purchase from the many sources of music, clearly beyond that of one conglomerate. Currently, the wholesale price of a digital track is approximately $0.68 a license which includes the artist and producer royalty, mechanical royalty, and revenue to label. (Remember that label revenue is *not* pure profit since the label funded the initial recording process along with any marketing effort.) The remainder of the $0.99 is distributed among the service provider, the distribution affiliate, credit card fees, and bandwidth costs.

Digital Revenues

As consumership of music continues to evolve, so does that of distribution of music. Current trends reveal that younger consumers look to purchase tracks, making the classic album less attractive to this buyer. The digital download arena is where these consumers are satiating their needs. In 2007, digital track sales grew 45% to a year-end sales total of 844 million. And the trend is continuing with the first quarter of 2008 showing an increase of 29% over the same time period last year. In 2007, digital album sales accounted for 10% of all album sales with the first quarter of 2008 showing a stronger trend of 15% of all album sales being digital. So, approximately 85% of albums and 78% of overall music is still purchased in the classic brick-and-mortar environment, making the need for physical distribution a continuing viable business entity.

Future Trends

It's but a snapshot of what 2008 might produce, but the first quarter is showing some interesting signs, with overall music purchases being up 6%. This includes albums, singles, digital and music video representing 431 million music purchases in 2008, as compared to 405 million purchases in 2007. Overall album, including T.E.A. are down 4% and album sales without T.E.A. are down 11% which includes all stratas except non-traditional, which is up 20%, but this increase does not make up for the other losses.

To date, Soulja Boy's Tell'em.com album has SoundScanned over 850,000 albums generating estimated revenues for the label in excess of over $8.5 million dollars. His single "Crank That" has digital download sales of 3.3 million at $0.99 a pop, with ringtone sales of over 2 million sold at an estimated $2.49 a download, generating over $8 million dollars in revenue. To look at the money, the single has generated nearly the same amount of revenue as the album, which is a new equation and business model for record labels and how they aggregate their assets.

DISTRIBUTION VALUE

Several distribution companies are exploring ways to add value to the conglomerate equation. Creating distribution-specific marketing campaigns with non-entertainment product lines helps validate distribution's existence, while hopefully enhancing the bottom line. Marketing efforts such as on-pack CDs with cereal, greeting card promotions, and ringtone services add to the branding of the participating artists, while increasing overall revenue through licensing and/or sales of primary items.

The ultimate value for today's distribution companies is that of consolidator and aggregator. Distributors can consolidate labels to create leverage points within retail. To gain positioning in the retail environment, one must have marketing muscle, and by using the collective power of their various label's talent, the entire company can raise its market share by coattailing on the larger releases in the family.

THE MUSIC RETAIL ENVIRONMENT

Marketing and the Music Retail Environment

The four Ps of product, promotion, price, and placement converge within the music retail environment—this being the last stop prior to music being purchased. This environment is designed to aid consumers in making their purchasing decision. This decision can be influenced in a number of ways, depending on the consumer. For example: "Does the store have hard-to-find releases?"; "Do they have the lowest prices?"; "Is it easy for consumers to find what they're looking for?"; "Does the store have good customer service and knowledgeable employees?" These questions should be answered, in one way or another, within the confines of the retail environment through the use of the four Ps.

NARM

To assist music retailers in determining business strategies, companies look for current sales trends as well as educational and support networks. The National Association of Recording Merchandisers (NARM) is an organization conceived to be a central communicator of core business issues for the music retailing industry.

Founded in 1958, NARM is an industry trade group that serves the music retailing community in the areas of networking, advocacy, information, education, and promotion. Members include brick-and-mortar, online and "click-and-mortar" retailers, wholesalers, distributors, content suppliers (primarily music labels, but also video and video game suppliers), suppliers of related products and services, artist managers, consultants, marketers, and educators in the music business field.

Retail members, who operate 7000 storefronts that account for almost 85% of the music sold in the $12 billion U.S. music market, represent the big national and regional chains—from Anderson (Wal-Mart wholesaler), Best Buy, Borders, Circuit City, Handleman (Wal-Mart and Kmart wholesaler), Hastings Entertainment, Musicland, Newbury Comics, Target, TransWorld (FYE), Virgin—to the small, but influential "tastemaker" independent specialty stores. Additional members also include online retailers such as Amazon.com, eBay, and iTunes®.

Distributor and supplier members account for almost 90% of the music produced for the U.S. marketplace. This includes the four major music companies: Sony BMG, EMI, Universal Music and Video, and WEA, and many of the labels they represent. A few of the label members are Atlantic, Blue Note, Capitol, Capitol Nashville, Columbia, Curb Records, DreamWorks Records, Elektra Entertainment, Epic Records, Geffen Records, Hollywood Records, Interscope Records, The Island Def Jam Music Group, J Records, Jive Records, Mercury/MCA Nashville/Lost Highway, RCA Label Group, RCA Music Group, RCA Victor Group, Rhino, Sony Classical, Verve Music Group, Walt Disney Records, Warner Bros., Welk Music Group, and Wind-up Records. This list extends even further to include influential independent labels and distributors. For more information regarding NARM and activities, visit www.narm.com.

Retail Considerations

The four Ps are applicable as an outline for retailer considerations in doing business.

#1 P—PRODUCT

How does a retailer learn about new releases? Distribution companies are basically extensions of the labels that they represent. To sell music well, a distribution company needs to be armed with key selling points. This critical information is usually outlined in the marketing plan that is created at the record label level. Record labels spend much time "educating" their distributors about their new releases so that the distributors can sell and distribute their records effectively.

Distribution companies set up meetings with their *accounts*, meaning the retailers. At the retailer's office, the distribution company shares with the *buyer*—that is, the person in charge of purchasing product for the retail company—the new releases for a specific release date, as well as the marketing strategies and events that will enhance consumer awareness and create sales. Depending on the size of the retailer, buyers are usually categorized by genre or product type, such as the R&B/hip-hop buyers or the soundtrack buyer.

Purchasing Music for the Store

When making a purchase of product, the buyer will take into consideration several key marketing elements: radio airplay, media exposure, touring, cross-merchandising events, and most critical, previous sales history of an artist within the retailer's environment; for if a new artist, current trends within the genre and/or other similar artists are considered. On occasion, the record label representative will accompany the distribution company sales rep with the hopes of enhancing the knowledge of the buyer of the new release. The ultimate goal is to increase the purchasing decision, while creating marketing events inside the retailer's environment.

Most record labels, along with their distributors, have agreed on a forecast for a specific release. This forecast, or number of records predicted to sell, is based on similar components that retail buyers consider when purchasing product. Many labels use the following benchmarks when determining forecast:

Initial orders or IO: This number is the initial shipment of music that will be on retailers' shelves or in their inventory at release date.

90-day forecast: Most releases sell the majority of their records within the first 90 days.

Lifetime: Depending on the release, some companies look to this number as when the fiscal year of the release ends, and the release will then rollover into a catalog title. But on occasion, a hit release will predicate that forecasting for that title continues, since sales are still brisk.

Store Types

A music store's target market or consumer generally dictates what kind of retailer it will be. To attract consumers interested in independent music, or to attract folks who are always looking for a bargain, determines the parameters in which a store operates. Music retailers have traditionally been segmented into the following profiles:

Independent music retailers cater to a consumer looking for a specific genre or lifestyle of music. Generally, these types of stores get their music from one-stops. Independent stores are locally or regionally owned and operated, with one or just a few stores under one ownership.

The *Mom & Pop* retailer is usually a one-store operation that is owned and operated by the same person. This owner is involved with every aspect of running the business and tends to be very passionate about the particular style of music that the store sells. This passion can be interpreted as being an expert in the knowledge of the genre and can be a unique resource for the consumer looking for the obscure release. Mom & Pop store owners tend to have a personal relationship with their customer base, knowing musical preferences, and keeping the customers informed about upcoming releases and events.

Alternative music stores profile very similarly to Mom & Pop stores, but with the exception that they tend to be lifestyle oriented. An electronic music retailer may have many hard-to-find releases along with hardware offerings such as turntable and mixing boards.

Chain stores tend to attract music purchasers who are looking for deep selection of releases along with assistance from employees who have strong product knowledge. These stores are often in malls and cater to a broad spectrum of purchasers.

These stores have been studied and replicated so that entering any store with the same name in any location feels very similar. Often, they have the new major releases upfront with many related items for sale, such as blank media and entertainment magazines. Chain stores traditionally buy their music inventory directly from music distributors, with warehousing and price stickering occurring in a central location. Some examples of chain stores are FYE, Hastings, and Musicland/Sam Goody.

Electronic Superstores do not make the bulk of their profits from the sale of music. But rather, use music as an attraction to bring consumers into their store environments. By using loss leader pricing strategies, these stores often sell new releases for less than they purchased them, but for a limited time. Meanwhile, they have created traffic to the store to make money from the sale of all the other items offered such as electronics, computers, televisions, and so forth. Electronic Superstores also buy their music inventory directly from music distributors, with warehousing and price stickering occurring in a central location. Some examples of electronic superstores are Best Buy and Circuit City.

Mass Merchants use the sale of music as event marketing for their stores. Each week, a new release brings customers back to their aisles with the notion that they will purchase something else while there. There is little profit in the sale of music for the mass merchant, but the offering of music is looked at as a service to customers. Often, mass merchants use rack jobbers to supply and maintain music for their stores. It is the rack jobber who initially purchases the music for the mass merchant environment. Some examples of mass merchants are Wal-Mart and Kmart.

Sources of Music

Figure 10.8 shows the basic flow of music as it reaches the consumer level. Recognize that one-stops' primary business is servicing independent record stores, but that they also do what is called *fill-in* business for all music retailers.

Internet Marketing and "Non-traditional" Retailing

Until recently, it was the retailers who had the brand identities that were winning the Internet sales wars. Consumers went to their favorite retail store websites to browse and purchase music, meaning the actual CD that was to be delivered to the consumer's door. These well-known retailer sites left many start-up websites with unknown names with little traffic. Sites such as iTunes®, Napster™, MusicMatch, Rhapsody, and yes, Wal-Mart are all experiencing a high volume of downloads, with more sites coming online soon. Well-known online sites such as Amazon and CDBaby have also had an impact on retailing. Plus non-traditional retailers such as Starbucks and Nordstroms have seen great success at selling finished goods to a target market. At the end of 2007, 18% of all albums sold were consumed at a "non-traditional" retailer, the lion's share being digital downloads, followed by internet sales, then "non-traditional" outlet such as Starbucks, and finally venues and mail order.

#2 P—PROMOTION

Survey after survey indicates that music consumers find out about new music primarily via the radio. So it would seem that music retailers would use radio as their

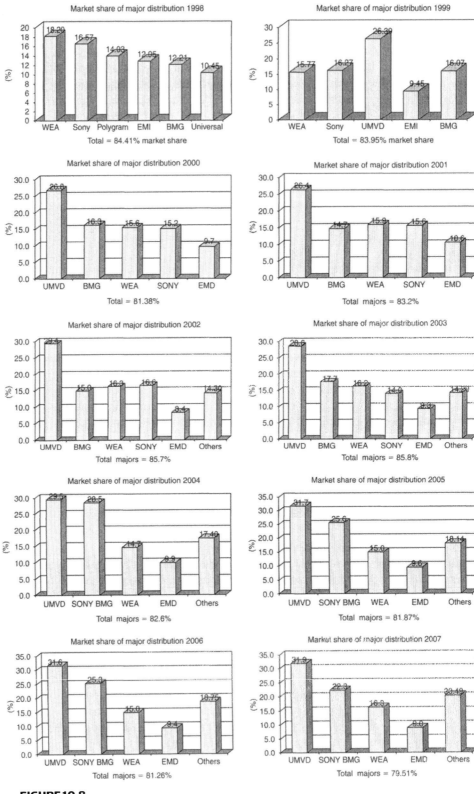

FIGURE 10.8

Distribution channels.

©2009 Amy Macy

primary advertising vehicle to promote their stores and product. But they do not. Cost considerations including reach and frequency of ads have caused music retailers to choose print advertising as their primary promotional activity. Consumers have been trained to look in Sunday circulars for sales and featured products on most any item. Major music retailers now use this advertising source as a way to announce new releases that are to street on the following Tuesday, along with other featured titles and sale product.

Once inside the store, promotional efforts to highlight different releases should aid consumers in purchasing decisions. These marketing devices often set the tone and culture of the store's environment. Whether it is listening stations near a coffee bar outlet, or oodles of posters hanging hodge-podge on the wall, a brief encounter within the store's confines will quickly identify the music that the store probably sells.

Featured titles within many retail environments are often dictated from the central buying office. As mentioned earlier, labels want and often do create marketing events that feature a specific title. This is coordinated via the retailer through an advertising vehicle called *co-operative advertising*. *Co-op advertising*, as it is known, is usually the exchange of money from the label to the retailer, so that a particular release will be featured. Following are examples of co-op advertising:

Pricing and positioning (P&P): P&P is when a title is sale priced and placed in a prominent area within the store.

End caps: Usually themed, this area is designated at the end of a row and features titles of a similar genre or idea.

Listening stations: Depending on the store, some releases are placed in an automatic digital feedback system where consumers can listen to almost any title within the store. Other listening stations may be less sophisticated, and may be as simple as using a free-standing CD player in a designated area. But all playback devices are giving consumers a chance to "test drive" the music before they buy it.

Point-of-purchase (POP) materials: Although many stores will say that they can use POP, including posters, flats, stand-ups, and so on, some retailers have advertising programs where labels can be guaranteed the use of such materials for a specific release.

Print advertising: A primary advertising vehicle, a label can secure a "mini" spot in a retailer's ad (a small picture of the CD cover art), which usually comes with sale pricing and featured positioning (P&P) in-store.

In-store event: Event marketing is a powerful tool in selling records. Creating an event where a hot artist is in-store and signing autographs of his or her newest release guarantees sales, while nurturing a strong relationship with the retailer.

The grid shown in Tables 10.3 and 10.4 show a sample forecasting and P&P planning tool used to predict IO and initial marketing campaign activities within a retailer's environment.

Table 10.3

Artist Name
Title
Selection Number

Account	% of Business	Target	Account Advertising P&P	Cost
Anderson Merchandisers	26.5	26,500		
Target	18.0	18,000		
Best Buy	15.0	15,000		
Alliance Entertainment B&N, Kmart, CCity	11.05	4,300		
Transworld	4.0	4,000		
Amazon	4.0	4,000		
Borders	3.0	3,000		
Hastings	1.25	1,250		
Costco	0.5	500		
TOP 10 accounts	83.25	83.25		
All others	16.75	16.750		
Total	100.0	100,000		

Rack jobbers such as Anderson and Handleman purchase nearly 30% of overall music. They supply music to mass merchants such as Wal-Mart and Kmart (Table 10.5).

Alliance is the largest one-stop, whose primary business is supplying music to independent music stores, but has also picked up retailers Barnes & Noble, Kmart, and Circuit City in recent days.

Figures are based on a major distributor's overall physical business with these actual accounts (Not Digital). Note that accounts purchase over 80% of all records, which makes them very important in the marketing process. For Target to *not* purchase a title would mean that over 18% of sales would be lost, based on these numbers. This does not mean that a record would sell 18% less over the lifetime of the release, but it does mean that consumers will have to find that title in another store other than Target.

What is not included in the Top Accounts is the digital business (downloads). For many labels, Apple's iTunes represents over 90% of their digital sales and if ranked within these above accounts, iTunes would probably be placed #2 with

Table 10.4	Sample Retail Chain P&P Costs and Programs						
Program	Program Includes	Mall Number of Stores	Rural (Strip Centers) Number of Stores	Media Number of Stores	Total Stores	Cost per Title ($)	Number of Titles
Best Seller	P&P on the Best Seller in-store position	500	250	75	825	22,500	20
New Release	P&P on the New Release in-store position	500	250	75	825	17,000	16
Music Video DVD	P&P on the Music DVD in-store position	500	250	75	825	2,700	8
Fast Forward	Positioning on the FF in-store position	500	250	75	825	11,000	12
Catalog Promotion	P&P on multiple prime in-store positions Supporting in-store signage Supporting promotional levers	500	250	75	825	*To be negotiated*	TBD
Premium Artist Package	P&P in all concepts TV in all concepts Insert (1 cut) (6.5 MM circulation) Magazine Direct Promotion Magazine RCC (monthly e-newsletter)	500	250	75	825	40,000 1,500 3,500 5,500 4,500	3
Total premium package price						*55,000*	

(continued)

Table 10.4

Program	Program Includes	Mall Number of Stores	Rural (Strip Centers) Number of Stores	Media Number of Stores	Total Stores	Cost per Title ($)	Number of Titles
Genre Endcaps							
POP1/RK1	P&P on endcap	500	250	75	825	13,500	12
POP1	P&P on endcap		250	75	325	3,500	2
RK1	P&P on endcap		250	75	325	3,500	2
POP2	P&P on endcap	350	200	75	725	10,000	8
POP3	P&P on endcap	250	75	75	400	7,000	8
RB1/RAP1	P&P on endcap	500	250	75	825	8,500	12
RB1	P&P on endcap		100	75	175	1,600	2
RAP1	P&P on endcap		250	75	325	3,500	2
URBAN 1	P&P on endcap	250	75	75	400	5,100	8
URBAN 2	P&P on endcap	150	41	76	267	3,300	8
OPT1 (open endcap)	P&P on endcap	350	200	75	725	5,000	8
OPT2 (open endcap)	P&P on endcap	250	75	75	400	3,000	8
CNJ1 (Classical New Age Jazz)	P&P on endcap	133		76	209	1,800	8

Table 10.5 Sample Rack Jobber P&P Programs and Costs	
Super feature: Sunday circular	$60,000 (1 month P&P)
Feature	$40,000 (1 month)
Genre features (1 month endcap) Urban	$5,000
Country	$5,000
Latin	$1,000
Pop: showboards (1 month)	$24,000
Urban: showboards (1 month)	$12,000
Country: showboards (1 month)	$20,000
CCM: showboards (1 month)	$5,000
Gospel: showboards (1 month)	$2,500
Various artists: showboards (1 month)	$10,000
Soundtrack: showboards (1 month)	$10,000
Rising stars developing artist (1 month)	$9,000
POD: position out of department (1 month endcap)	$5,000 $7,500
Urban direct ROTO (1 week)	$1,500 single cut $3,000 double cut
ROTO: Sunday circular (1 week)	$10,000 per cut

approximately 20% market share. These sales are not shipped product but are usually digital single sales that are accounted with TEA equivalency, making this percentage of business more difficult to calculate.

#3 P—PRICING

Although record labels set the suggested retail list price (SRLP) for a release, this is *not* what retailers are required to sell the product for. Most often, the SRLP sets the *wholesale price* or the cost to the retailer. In negotiating the order, the retailer may ask for a discount off the wholesale price. The retailer may also ask for additional *dating*, meaning that the retailer is asking for an extension on the payment due date. Each distributor has parameters in which this transaction may occur.

Generally, music product comes in box lots of 30 units. A retailer will receive a better price on product if it purchases in box lots. For example, a retailer wants to purchase 1200 units of a new release with a 10% discount and 30 days dating (Table 10.6).

Table 10.6

SRLP	$18.98	SRLP	$18.98	The 10% discount
Wholesale	$12.04	Wholesale	$12.04	saved the retailer
×	1,200 units	−10% discount	$1.20	$1,440
	$14,448	Total/unit	$10.84	
		×	1,200 units	
			$13,008	

Table 10.7

$18.98	SRLP
$12.04	Wholesale
Label	$5.00
Distribution	$1.80
Design/manufacturing	$1.00
Artist royalty	$1.50
Mechanical royalty	$0.80
Recording costs	$1.00
Marketing/promotion	$0.90
Retail profit	$3.00

Normally, the money due for this purchase would be received at the end of the following month that the record was released. However, with an extra 30 days dating, the due date is extended, giving the retailer a longer time frame in which to sell all the products. Adding extra dating is often a tactic of record labels that want retailers to take a chance on a new artist that may be slower to develop in the marketplace.

The price of product reflects the store's marketing strategy. The major electronic superstores look to music product as the magnet to get customers through their doors, which is why music prices in these environments are often lower than anywhere else. Often, these stores will sell music for less than they purchased it, called *loss leader* pricing. But these stores will also raise the price after a short period of time, usually within first 2 weeks after the street date.

Where the Money Goes

The consumer plunks down their money at the cash register to purchase a CD. Table 10.7 shows how that money is divided between all the invested parties.

If the retailer receives a 10% discount off the wholesale price, then the label gets a reduction of 3% and the distributor a reduction of 2%, while the retailer enjoys

Table 10.8 Where the Money Goes with 10% Discount to Retailer

Revenues from a $99 Download

Participant	Revenue (cents)
Label	47
Distribution affiliate	10
Artist	7
Producer	3
Music publisher	8
Service provider	17
Credit card fees	5
Bandwidth costs	2
	99

Source: Garrity (2003).

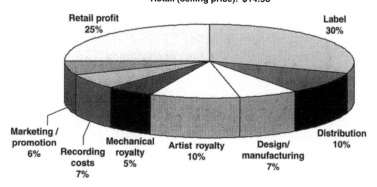

Where the money goes
SRLP: $18.98 Wholesale: $10.87 with 10% discount to retailer
Retail (selling price): $14.98

Retail profit 25%
Label 30%
Marketing / promotion 6%
Recording costs 7%
Mechanical royalty 5%
Artist royalty 10%
Design/ manufacturing 7%
Distribution 10%

FIGURE 10.9
Sample distribution pricing schedule.

a 25% *increase* in margin, from 20% of retail price to 25%, as shown in Table 10.8 (Figure 10.9).

#4 P—PLACE

Although promotional activities within the store include the *place*ment of product in designated areas, the choosing of a location or place of the actual store is paramount to the success of the business. Factors to consider are:

Location and visibility: Having built-in traffic helps a store attract more customers. Being in a mall or a high traffic area is ideal for the store selling mass-appeal products. Additionally, being visible, easy to find and access can help a store succeed.

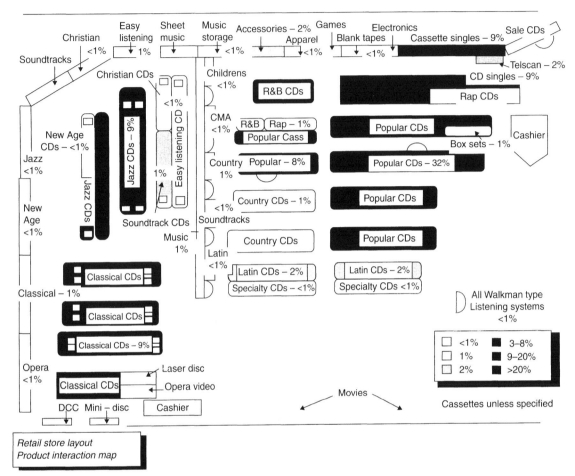

FIGURE 10.10
Retail store layout.

Competition: Depending on the store's marketing strategy, having competition nearby can detract from a store's success. Knowledge of area businesses is key to success.

Rent: Besides inventory and employee costs, rent is a big expense that can impact profitability. Again, depending on store strategy, some retailers build their own storefront, while others rent from existing outlets.

Use clauses: A store must be sensitive to non-compete use clauses that can restrict business. Some malls restrict the number of record retail outlets and/or electronic store outlets to help ensure success of existing tenants.

Image: Store image reflects corporate culture along with product for sale.

Floor plans: The store's culture and marketing strategy is best observed in the design and layout of the floor plan. Market research observes that the longer a person is in a store, the higher the probability that the person will purchase an item, based on a marketing concept called *time spent shopping (TSS)*. On that premise, some stores draw in their customers, creating interesting and interactive displays further back in the store. Sale items may be placed in the back, or in some cases, there may be multiple floors, all with an eye to keep the shopper in the store longer. Other stores may entice browsers with new releases right inside the front door. Others include the addition of coffee bars to boost TSS.

Genre placement, along with related items, helps define a store. Display placement, including interactive kiosks, listening stations, featured titles, and top-selling charts, helps consumers make purchasing decisions. Larger stores with broader product offerings couple music product with related items. The placement of these elements and traffic flow design can aid consumers and increase sales (Figure 10.10).

GLOSSARY

Big 4 These are the four music conglomerates that maintain a collective 85% market share of record sales: they are Universal, Sony BMG, Warner, and EMI.

Bricks and clicks The term given to a retailer who has physical stores and an online retail presence. See also **Click and Mortar**.

Brick and mortar The description given to physical store locations when compared to online shopping.

Box lot Purchases made in increments of what comes in full, sealed boxes receive a lower price. (For CDs with normal packaging, usually 30.)

Buyers Agents of retail chains who decide what products to purchase from the suppliers.

Chain stores A group of retail stores under one ownership and selling the same lines of merchandise. Because they purchase product in large quantities from centralized distribution centers, they can command big discounts from record manufacturers (compared to indie stores).

Click and Mortar The term given to a retailer who has physical stores and an online retail presence. See also **Bricks and Clicks**.

Computerized ordering process An inventory management system that tracks the sale of product and automatically reorders when inventories fall below a preset level. Reordering is done through an electronic data interchange (EDI) connected to the supplier.

Co-op advertising A co-operative advertising effort by two or more companies sharing in the costs and responsibilities. A common example is where a record label and a record retailer work together to run ads in local newspapers touting the availability of new releases at the retailer's locations.

Discount and dating The manufacturer offers a discount on orders and allows for delayed payment. It is used as an incentive to increase orders.

Distribution A company that distributes products to retailers. This can be an independent distributor handling products for indie labels or a major record company that distributes its own products and that of others through its branch system.

©2009 Amy Macy

Drop ship   Shipping product quickly and directly to a retail store without going through the normal distribution system.

Economies of scale   Producing in large volume often generates economies of scale—the per-unit cost of something goes down with volume. Fixed costs are spread over more units lowering the average cost per unit and offering a competitive price and margin advantage.

Electronic data interchange (EDI)   The inter-firm computer-to-computer transfer of information, as between or among retailers, wholesalers, and manufacturers. Used for automated reordering.

Electronic superstores   Large chain stores such as Circuit City and Best Buy that sell recorded music and videos, in addition to electronic hardware.

End cap   In retail merchandising, a display rack or shelf at the end of a store aisle; a prime store location for stocking product.

Fill-in   One-stop music distributors supply product to mass merchants and retailers who have run out of a specific title by "filling in" the hole of inventory for that release.

Floor designs   A store layout designed to facilitate store traffic to increase the amount of time spent shopping (TSS).

Free goods   Saleable goods offered to retailers at no cost as an incentive to purchase additional products.

Indie stores   Business entities of a single proprietorship or partnership servicing a smaller music consumer base of usually one or two stores (sometimes known as *mom & pop* stores).

Inventory management   The process of acquiring and maintaining a proper assortment of merchandise while keeping ordering, shipping, handling, and other related costs in check.

Listening station   A device in retail stores allowing the customer to sample music for sale in the store. Usually the devices have headphones and may be free standing or grouped together in a designated section of the store.

Loose   The pricing scheme for product sold individually or in increments smaller than a sealed box.

Loss leader pricing   The featuring of items priced below cost or at relatively low prices to attract customers to the seller's place of business.

Margin   The percentage of revenues left over to cover expenses as well as account for profitability.

Markup   The percentage of increase from wholesale price to retail price.

Mass merchants   Large discount chain stores that sell a variety of products in all categories, for example, Wal-Mart and Target.

Min/max systems A store may have 10 units on hand, which is considered the ideal *maximum* number the store should carry. The ideal *minimum* number may be 4 units. If the store sells 7 units and drops below the ideal inventory number of 4, as set in the computer, the store's inventory management system will automatically generate a reorder for that title, up to the maximum number.

National Association of Recording Merchandiser (NARM) The organization of record retailers, wholesalers, distributors, and labels.

One-Stop A record wholesaler that stocks product from many different labels and distributors for resale to retailers, rack jobbers, and juke box operators. The prime source of product for small mom & pop retailers.

Point-of-purchase (POP) A marketing technique used to stimulate impulse sales in the store. POP materials are visually positioned to attract customer attention and may include displays, posters, bin cards, banners, window displays, and so forth.

Point-of-sale (POS) Where the sale is entered into registers. Origination of information for tracking sales, and so on.

Price and positioning (P&P) When a title is sale priced and placed in a prominent area within the store.

Pricing strategies A key element in marketing, whereby the price of a product is set to generate the most sales at optimum profits.

Rack jobber A company that supplies records, cassettes, and CDs to department stores, discount chains, and other outlets and services (racks) their record departments with the right music mix.

Returns Products that do not sell within a reasonable amount of time and are returned to the manufacturer for a refund or credit.

Sales Book Distribution companies compile all their releases for a specific street date into a "sales book," which contains one sheet for each title that outlines the marketing efforts.

Sales forecast An estimate of the dollar or unit sales for a specified future period under a proposed marketing plan or program.

Sell-through Once a title has been released, labels and distributors want to minimize returns and "sell-through" as much inventory as possible.

Shrinkage The loss of inventory through shoplifting and employee theft.

Solicitation period The sales process of specific titles occurs during a period called *solicitation*. All titles streeting on a particular date are placed in a solicitation book, where details of the release are described.

Source tagging The process of using electronic security tags embedded in a product's packaging.

Theft protection Systems in place to reduce shoplifting and employee theft in retail stores. These systems may include electronic surveillance.

Time spent shopping (TSS)   A measure of how long a customer spends in the store.

Turn   The rate that inventory is sold through, usually expressed in number of units sold per year/inventory capacity on the floor.

Vertical integration   The expansion of a business by acquiring or developing businesses engaged in earlier or later stages of marketing a product.

Universal Product Code (UPC)   The bar codes that are used in inventory management and are scanned when product is sold.

Wholesale   The price paid by the retailer to purchase goods.

REFERENCES

Garrity, B. (2003). "Seeking Profits at 99¢," *Billboard*, July 12.
Knab, C. (2001). www.musicdish.com/mag/index.php3?id = 3357. The Distributor One Sheet, March, 25.

FURTHER READING

This adapted chapter is taken in part from Hutchinson et al. (2005). *Record Label Marketing*. Focal Press.
Nielsen SoundScan State of the Industry 2007–2008.
Weatherson, J. (2004). President of Ventura Distribution and former Executive Vice President of Universal Music Distribution, from personal interview, July 2004.

SECTION D
Doing It Yourself

Chapter 11
The DIY Approach

Russ Hepworth-Sawyer

In This Chapter

In this section of the book, we look at how you can take matters into your own hands and record the music yourself. This is affectionately known as the DIY approach. In this chapter we introduce the notion of doing it yourself and where to look for more information.

INTRODUCTION

Many more musicians are now investigating the possibility of becoming self-sufficient by taking the production process into their own hands. At some point a decision may be made to start a small recording "business" at home because you want to record music in a certain way, or it may be simply to save on costs.

Along with development and greater proliferation of ever-cheaper, higher quality audio equipment, came a growing number of recording enthusiasts. Musicians were able to begin recording their material at home and increasingly achieving professional standard results.

For many, this remains the only way in which their music can be recorded and brought to a physical format. With the expansion of digital delivery over the Internet using online stores such as iTunes® and the download chart has meant that even physical delivery is not always necessary to distribute music and be successful.

It is possible now for the do-it-yourself (DIY) artist to follow the whole production process from start to finish in the confines of their own home without the need to hire the professionals unless a large-scale physical release is wanted.

EQUIPMENT

Recording equipment is much cheaper than 15 years ago when named professional mixing consoles were much more expensive than the average house here in the UK (by today's prices!), and the same again for a decent professional multitrack tape machine. However, to get set up today can still take quite a lot of money, so decisions need to be taken after careful consideration of the work you're likely to be doing.

Researching and buying equipment should be an exciting and rewarding experience. However, there are so many choices that just settling on one piece of gear compared to another is a difficult task. Again, when you're starting out and money is tight, it is difficult to buy all the gear you might want or need straight away, so compromises must be made. Decisions need to be made about quantity or quality. Compromising on those expensive, yet excellent, monitors and settling for something else can free up the remaining budget to get a wider range of affordable microphones, although that is a choice you will need to make.

The solution to this particular problem might be to make it modular! For example, if you cannot afford those large monitors by that top of the range brand, consider buying one of their smaller, more affordable pairs, which a subwoofer can be added to at a later date. Think how each purchase can get you a little closer to where you would like to be. Buy one good microphone now, if you can afford it, with a view to getting another of the same model later to make a stereo pair, although not what we would call a matched stereo pair, with the common consecutive serial numbers.

As can be seen by the example above, many decisions need to be made about the equipment required. Below we look at some of the equipment you may wish to include in your studio environment. Please also refer to the further reading list at the end for sources of information.

The Computer

The biggest change from 15 years ago is that most people now own a Windows-based PC for their home use. A computer (whether Apple or PC) is the hub of most digital studios installed with the appropriate software. Before taking the family machine into the spare room to make a studio, consider the following aspects.

Computers are all the same, but they vary so much. The variations are important when you ask a lot of the machine's power. For example, a typical family PC should be absolutely adequate for the most demanding of Internet browsing and word processing. Introduce 16 tracks of audio at 44.1 kHz and 24 bit with a few plug-ins here and there and it might begin to grind a little. Not all machines are made equally. If the family machine is all that is available, bear in mind that your progressive rock opera with five orchestras playing simultaneously (one for each full range speaker in a 5.1 surround sound system) might have to be put on pause.

If you're likely to buy a PC machine especially for the purpose of making music, then fortunately there are lots and lots of options. There are specialist dealers and manufacturers that make machines for the purpose of audio and can handle most projects (apart from perhaps the most demanding of 5.1 progressive rock operas!).

Many studios choose to work on the Apple Mac platform, despite solid Windows-based versions of some of the main software. With Apple computers there is no need to specify which one has been built for audio as they do not distinguish between tasks. A Mac is a Mac, simply choose the computer you wish for most audio projects, or the most powerful one you can find before embarking on any large-scale projects.

The Software (Digital Audio Workstation)

As for software choices, there are simply a proliferation of them ranging from very affordable and powerful audio editors such as Adobe Audition and even the free Audacity, all the way through to specific professional standard systems such as ProTools.

What you choose will be up to you, but at the time of writing, for recording and producing music, there are a handful of main contenders. The one at the front of the queue for recording is Digidesign's ProTools which has two flavors of system. The first is the HD system that uses installed digital signal processing (DSP) cards to do the number crunching when adding a plug-in. This is the professional system and is in constant use around the world. ProTools also comes in a LE form that instead uses the computer's processing power to "number crunch" the DSP. ProTools operates on both Mac and PC.

Another main contender is Apple Logic Studio. Logic is an excellent package that includes Logic Pro, their audio and MIDI sequencer, which has a huge following around the world. It is a very powerful system and comes bundled with pretty much everything a recording musician would need, even a live performance program called

Main Stage and a mastering application called Waveburner. Logic also comes in a lighter version called Logic Express. Logic only operates on the Mac platform.

Cubase, by Steinberg, has been one of the most popular MIDI sequencers over the years. It was one of the first pieces of software to introduce the idea of plug-ins with "VST" (Virtual Studio Technology). Cubase continues to be extremely popular on both the Mac and PC platforms.

Digital Performer, by Mark of the Unicorn (MOTU), has a large following also. This Mac application has some exceptional users working both in the recording industry as well as composers in the film and television industry. It has some excellent plug-ins such as the Masterworks EQ and Compressor. MOTU also makes some excellent and well-respected audio and MIDI interfaces.

Reason, by Propellerhead Software, although not for recording, has been extremely popular over the years for it offers a complete environment for making electronic music with some excellent synthesizers, samplers and effects, in addition to a mixer and sequencer. Reason can be connected invisibly to most other Digital Audio Workstations (DAWs) using ReWire, a protocol which allows for MIDI and Audio signals to be transferred between two applications running on the same computer.

There are, of course, many more systems not mentioned here. Take a look at any of the recording magazines in your area to find out who the manufacturers are and any reviews that might be available.

DECISIONS...DECISIONS...DECISIONS

Taking the plunge and spending out for equipment is always tricky, and to make a choice of the environment in which you're going to record is always going to be a difficult decision. Making that decision will be completely up to you but there are some potential points to consider.

Your genre of music can help you make the right decision. For example, the systems mentioned in the previous section have grown up from slightly different places in the audio market. For example, Cubase was always simply a MIDI sequencer, as was Logic (as Notator), whereas ProTools was always predominately an audio package which once had little extensive need for MIDI.

To some extent the strengths of these systems are still in their historical starting points. For example, ProTools is the undisputed market leader when it comes to recording audio in the music industry. For MIDI-based sequencing Logic and Cubase seemingly lead the way.

However, convergence has now meant that someone wishing to program MIDI-based work within ProTools most certainly can, and vice versa, Logic can quite easily handle your multitrack project.

Therefore, if you're a folk artist and wish to start your tracking at home and move on to a mix engineer's studio, then ProTools might be the right system for you to look to. Whereas, if your work is predominately electronic, then a system such as Logic, Cubase or Digital Performer with Reason ReWired might be a better choice.

Try as many equipment demos as possible. Speak to as many users as you can, and there are some exceptional web forums out there with plenty of information. Consider your genre and what your peers, and the genre's producers and engineers, are using.

ADDITIONAL PLUG-INS AND SOFTWARE INSTRUMENTS

Plug-ins are the software equivalent of adding a reverb unit to your rack of effects (see the section on Outboard). They allow you to insert an effects or dynamics processor into your DAW-mixing environment without leaving the digital domain. Software instruments are also plug-ins but allow you to insert a synthesizer or sampler into your DAW.

These "devices" can often be third party as they allow you to gain the sound and feel of some of the older, standard coveted equipment. For example, you can purchase a software equivalent of famous EQs, compressors, limiters, and so on. The other major benefit to plug-ins is that once you have it installed on your system, you can use as many instances of it as your processor can manage, which is something you cannot do with your racked-up famous EQ box.

There are a startling amount of plug-ins and software instruments available these days. All manner of replications and new inventions are offered in the software environment. It is worth considering your position at this point. Of course we'd all love to have a wealth of plug-ins but they cost lots of money. Can you get along with the plug-ins that come with the DAW, as most of them are shipped with an excellent range?

The Interface

When recording into a computer you will most likely need an audio interface. If you are working with ProTools, you will have an interface provided, but if you choose one of the other software providers, you will have a wider range of interfaces to select from.

Interfaces come in many different forms from simply USB connected devices, to FireWire and then PCI-Express devices. The simple USB devices generally provide two to four inputs and a selection of outputs. Some FireWire devices can extend this considerably to eight inputs or more, with a similar number of outputs for example. For the more professional setup requiring many more inputs, a PCI-based system will be required allowing many more channels.

Choosing the interface is an important part in the process as this is where a lot of the sound quality is achieved. The job of converting the analog waveform into digital and back again is important as it is at this stage where the quality of the audio is maintained and reproduced. An affordable interface can no-doubt provide excellent results, but professionals spend a great deal of money choosing the right convertors for their system to ensure that the highest sound quality is achieved.

Whatever interface you choose, ensure it has the capability to record the inputs and outputs you need. For example, does it have enough XLR mic inputs for the drum kit? How many microphones can be phantom powered (see below)? What provision for line inputs are there? Are there any Hi-Z (high impedance) instrument inputs for plugging your guitars straight in?

Another question at this stage to consider is whether you're going to use a mixing console or not. A mixer can of course provide the microphone inputs and phantom power needed for some microphones. If a console is going to be used it will have the ability to handle the volume control for your monitors, talkback for your band when you're at the controls. If no console is to be used, then you may wish to ensure that the audio interface can handle these duties.

Microphones

There are two main popular types of microphones to consider when recording: Condenser and Dynamic microphones.

Condenser microphones are the slightly more expensive and delicate of the two. They are more sensitive generally and have a wide frequency response. They are excellent for quiet and loud sounds and are frequently used on all types of recordings.

Dynamic microphones possess a slightly limited frequency response to the condenser microphone, are more robust and are less sensitive. These properties allow these microphones to be used in differing situations. Dynamic microphones can be used on very loud instruments. Because they are less sensitive they capture less spill from around them while increasing separation.

Choosing which microphones are best for you is outside the remit of this book as this will be a personal choice. However, there are some points to consider before thinking about your choice. All microphones generally use XLR connectors and it is worth ensuring that your input to your recording device (presumably computer) allows these. Additionally if a condenser mic is to be used ensure that it provides something called Phantom Power. Condenser mics require this to enable the microphone capsule to operate and to power the internal microphone pre-amplifier.

Many excellent publications exist discussing the theory and use of microphones, which can be found in the further reading list.

Monitors

Monitors, or loudspeakers, are very important in the studio. As you may have seen in many pictures of studios there are at least two sets of them if not more. This is not obligatory, but can give the engineer different perspectives on the same mix. One set will be the far-field (a long way away and perhaps soffit mounted in the wall) whilst the other will be on the meter bridge known as near-field monitors. As you set up it is unlikely that you will have either the money or space to get an enormous pair of far-field monitors, especially if they should be soffit mounted in the wall. Therefore you will need to look toward a smaller size.

You could use your hi-fi speakers if you know them and are happy working with them. For many, they get to know a set of speakers and then wish to stick with them. So many engineers and studios still keep the Yamaha NS10M Studio monitors in their rooms (the ones with the white woofers always on the meter bridge) as they have become part of the studio furniture for so many. This is not because they are excellent monitors—some would say "far from it"; they just offer a dimension to mix assessment that has provided popular to so many engineers. The age old adage states "if it sounds good on NS10s—it is good."

Sadly NS10s are no longer manufactured, but the good news is that the near-field monitor market is bustling with different models on show. There are so many to choose from now. There are some excellent models for a reasonable price and it would be wise spending a considerable amount of time listening and deciding what monitors are best for you … and your environment! The room in which they are put and where they are placed within that room are the make or break deal whether they'll perform as you expect. See the acoustics and design sections later in this chapter.

Mixing Consoles and Control Surfaces

You may already own a mixing console, or board, or desk and it might be prudent to integrate this into your setup. Alternatively you may wish to use a desk to mix on. So much of the desk's functionality can be carried out within the Digital Audio Workstation that it often becomes a device with two channels brought up to handle the stereo output from the computer. However, a desk can offer many more benefits.

As you gather your studio together, it is unlikely that a console will be required or necessary, but as your setup grows you may wish to include one. There are many engineers who prefer to mix on a console for the visual representation of the mix laid out before them. A modern approach to the visual representation appears to be in the popularity of the control surface. Control surfaces offer a physical representation of the software environment in which you're working and provide the tactile aspect of being a sound engineer.

Other engineers prefer to mix on a console because of the analog summing that it provides. Summing refers to how each signal is merged together to give one stereo representation of the music. Some believe that analog consoles can provide superior sound quality through the EQ or other facilities it provides.

Your reasons for wanting a desk might be varied, but consider the additional benefits they provide. Mixing consoles provide a volume control, talkback and other interesting features that are worth taking into account. For more details on mixing consoles please refer to the further reading section at the end of this chapter.

Outboard

Before the advent of DAWs, when mixing consoles and multitrack tape machines dominated, all audio processing took place in devices called outboard. Outboard is a catch-all word for all the equipment in the racks away from the console. These include dynamics processors (expanders, gates, compressors, limiters) and effects processors (reverberation, chorus, delay, etc.).

As time has gone on and the DAW is installed pretty much everywhere, the need to outboard has decreased over the years, although there are many who look for that special ingredient an analog compressor or limiter provides, perhaps containing a valve or two for that "warmth" factor. A new revived market on high quality outboard still exists and provides some excellent options. Similarly the high-end effects processor market is still sought after.

General Kit

There will always be the need for a pair of headphones for you to listen to when you're tracking, but also perhaps for listening to as you record or mix. Choosing the right headphones is important as they will need to be of a closed variety for tracking as this prevents sound leaking out of them and spilling into your microphone and recording, also ensuring they can endure the power necessary for the most demanding of drummers to hear what the rest of the band sound like. Shop around and look at what many studios use.

Cabling is always something that people seem to forget about until the last minute. Actually when it is considered, some forget to give ample enough budget to it. Cabling is very important as any degradation in your signal can have devastating issues across many parts of your studio. Spend a little time researching about cabling and where it matters.

Microphone stands are again something that some forget until the last minute and try to save money. Just think that one day you might place your pride and joy—an expensive condenser microphone—as an overhead for it to slowly sag and either hit the drummer on the head or fall to the floor tom. Not only will it be damaging to the mic and your monitors, but also the drummer's floor tom skin might be broken. Take some time to do a little research.

Making the Right Choices

Making equipment choices can be difficult, and a certain degree of fact finding is required so as to make the best purchase. Many recording magazines offer excellent reviews and articles on equipment, some even provide buyers guides to help you through the sheer range available. Speaking to professionals and seeing what they have to say on some of the main forums is another way to make savvy purchases. Again consider the genre of music you're making or recording and seek out the people who record it and their views. It is very easy to get tempted by the wealth of gear and views on gear out there. If a folk album is all you're trying to make, then consider the equipment you need to make that album.

This section has only been intended as an introduction to the basic main elements of the studio. For more information on recording equipment please refer to the excellent sources shown in the further reading list at the end of this chapter.

STUDIO CONSIDERATIONS

Recording at home is not always as simple as installing some software and taking over the spare room. There are a number of other considerations which should be looked at. Before obtaining and installing your equipment, it is worth bearing in mind where exactly you will be recording.

Space

In the professional recording industry, huge amounts of money are lavished on very nice looking spaces for recording. They not only look expensive, they should sound expensive too. Creating the right "vibe" is important in capturing the correct performance, but also it is imperative that the sound is accurate when monitoring your recordings.

NOISE

The biggest problem you're likely to face is noise when recording at home; noise that others produce will make an appearance on your recordings, and the noise you and your drummer emit as you're tracking that last track late one Monday evening.

It is worth explaining at this point the difference between soundproofing and acoustic treatment, which can so often be misinterpreted. Soundproofing is an expensive, heavy and messy business. To stop high sound pressure levels, such as that caused by a drum kit or a cranked up guitar amp, requires some pretty permanent solutions. These solutions usually include building rooms within rooms, floating floors, air locking doors and some attenuating ventilation or air conditioning. On the other hand, acoustic treatment is much simpler and cheaper to achieve, as its purpose is to manage the sound within a particular space so that the listener's experience is exacting. In an ideal world any studio space should be a perfect marriage between soundproofing and acoustic treatment.

ACOUSTICS

Acoustically treating your recording spaces is very important if you're to capture sonically excellent takes and be able to monitor them without coloration or inaccuracy. Doing this level of work at home may not always be necessary, but is certainly a consideration when working. If you're recording drums at home, then a reverberant room may work absolutely fine for some music, but might be the wrong things for another. Poor monitoring environments could mean that mixes are poorly translating when listened to in other places (bass heavy or bass light for example).

There are plenty of solutions and DIY approaches to acoustically treat rooms. Some simple rules and methods exist which can be easily researched. These are not addressed in this book, but many articles and publications exist and some are listed in the further reading section at the end of this chapter.

DESIGN

Treatment of the acoustics is one thing, but the design of the furniture and your seating position in relation to it and the room will make a huge difference to the sound. Designing a room that looks good on paper is not always the best sonic answer. Moving your monitors by a couple of inches back and forth can make an exceptional difference how it works with the room. Thus, it is often advised to do some listening in a room without the equipment or furniture to locate the best place for you and your monitors. With this determined then the furniture and equipment can be installed.

The ergonomics of your studio environment are also very important. Ergonomics refer to the position of your equipment, your seating position and the comfort and efficiency this provides. Studio designers work with specialist furniture makers to ensure that the studio is ergonomic and perfectly comfortable to use for extended periods of time.

Take a peek at the websites of some of the major studio installers to see the facilities they have installed lately for ideas on how to tackle your space. Consider the look and feel of your room to ensure it is not only sonically solid, but is ergonomically sound and somewhat creative to work within.

Knowledge

Recording engineers spend a lifetime learning and perfecting their art and whilst you may have similar equipment, this is of no use without some of the knowledge. There are so many excellent books and magazines available to you to get up to speed with techniques and traditions within the industry. These are listed in the further reading section at the end of this chapter.

Spending time just reading one or two of these books will immediately expand your knowledge of microphone technique, sound engineering skills, effects and dynamics processing and a whole multitude of skills. Reading these will immediately improve your ability to capture excellent recordings. Subscribing to one of the music technology magazines published in your area will offer a wealth of information about equipment and techniques.

In most areas there are likely to be recording courses that you can attend, whether part or full time. Courses such as these will allow you to gem up on your theory, or experience some recording practices or equipment unbeknownst to you. There are some distance learning courses if there are none in your area.

Objectivity

It is quite possible to be able to do all the work yourself from start to finish. However, there will come times when objectivity appears to escape you. There are many times in the process where an impartial ear will be useful. Whether that be at the writing stage, pre-production or mixing stages, some objectivity is productive.

Again, whilst it is perfectly possible to go all the way through the process solely, it would be advisable for a mastering engineer to finish the process for you, ensuring that the product is in line with everything else out there and will work on the systems it will be played on. The added advantage here is you can hear things on their system that you had not heard before on your monitors.

The following chapters examine some of the key skills you will need when recording, starting with "The Recording Process: A Guide" in which we examine the ways in which recording can take place. Following this is "Evaluating

Sound Quality" which is an excellent guide to problems in sound quality and how to possibly address them. Finally is a short chapter with some hints and tips as you go through the production process which is expanded on the www.demo2delivery.com website.

FURTHER READING

Collins, M. (2006). *Pro Tools Le and M-Powered: The Complete Guide*. Focal Press, Oxford.

Cousins, M. & Hepworth-Sawyer, R. (2008). *Logic Pro 8: Audio and Music Production*. Focal Press, Oxford.

Critch, T. (2005). *Recording Tips for Engineers: For Cleaner, Brighter Tracks*. Focal Press, Oxford.

Eargle, J. (2004). *The Microphone Book*. Focal Press, Oxford.

Harrison, A. (2008). *The Music Business: The Essential Guide to the Law and the Deals*. Virgin Books, London.

Huber, D. & Runstein, R. (2005). *Modern Recording Techniques*. Focal Press, Oxford.

Izhaki, R. (2008). *Mixing Audio*. Focal Press, Oxford.

Newell, P. (2003). *Recording Studio Design*. Focal Press, Oxford.

Rumsey, F. & McCormick, T. (2005). *Sound and Recording: An Introduction*. Focal Press, Oxford.

Stavrou, M. (2003). *Mixing with Your Mind*. Flux Research, Australia.

Stone, C. (2000). *Audio Recording for Profit: The Sound of Money*. Focal Press, Oxford.

www.apple.com

www.digidesign.com—The makers of ProTools.

www.merging.com—The makers of Pyramix DAW.

www.apple.com/logic—Logic Studio by Apple.

www.steinberg.de—The makers of Cubase and Nuendo.

audacity.sourceforge.net—Audacity Audio Editor.

www.propellerheads.se—The makers of Reason and ReWire.

www.motu.com—The makers of Digital Performer and a host of well respected interfaces.

www.acoustipro.com—Makers of excellent acoustic optimizing products.

www.auralex.com—Makers of foam-based acoustic treatment.

www.realtraps.com—Makers of bass traps and other acoustic products.

modrec.com—The website to accompany Huber and Runstein's Modern Recording Techniques.

www.aaa-design.com—Recording architecture.

www.demo2delivery.com—This book's website on which more details can be found.

Chapter 12

The Recording Process: A Guide

Bruce and Jenny Bartlett

In This Chapter

In this chapter, Bruce and Jenny Bartlett discuss the options available to you to begin recording and enter into the production process yourself by recording that initial demo, or if you are accomplished enough, an album.

INTRODUCTION

Welcome to the brave new world of 21st century recording! The digital technologies of the past few years have given us possibilities undreamed of just 10 years ago. Thanks to the shift from analog to digital technology, the excitement and satisfaction of recording are accessible to more people than ever before. It used to take a whole roomful—or truckful—of expensive equipment to produce a good recording.

But the new generation of smaller, cheaper gear means you may be able to tuck a studio into a corner of your bedroom or the back seat of a car. As a result, many more people are involved in the process of recording. As musicians are recording their own albums more these days, this chapter both discusses how you might wish to get started, including the key ways of looking at the recording process, as well as how an engineer might record your music.

Recording engineers are key players. Their skills help artists realize their visions in sound. The engineer's miking techniques capture the vibrancy of the performance, whether it's the shimmering overtones of a string quartet or the sonic assault of an electric blues band. Their work in the studio—adding effects, tweaking levels, etc. will take the raw material of the performance and shape and blend it into a polished musical statement. Mastering the technology, becoming fluent with the audio tools at hand, they will produce exciting recordings that will delight clients and give you an excellent product.

TYPES OF RECORDING

Let's get started. Currently there are six main ways to record music:

1. Live stereo recording
2. Live-mix recording
3. Multitrack recorder and mixer
4. Stand-alone recorder–mixer
5. Computer Digital Audio Workstation (DAW)
6. MIDI sequencing

Live Stereo Recording

Record with a stereo microphone or two microphones into a recorder. This method is most commonly used to record an orchestra, symphonic band, pipe organ, small ensemble, quartet, or soloist. The microphones pick up the overall sound of the instruments and the concert hall acoustics from several feet away. You might use this minimalist technique to record a folk group, rock group, or acoustic jazz group in a good sounding room.

Figure 12.1 shows the stages of this method—the links in the recording chain. From left to right:

1. The musical instruments or voices make sound waves.
2. The sound waves travel through the air and bounce or reflect off the walls, ceiling, and floor of the concert hall. These reflections add a pleasing sense of spaciousness.
3. The sound waves from the instruments and the room reach the microphones, which convert the sound into electrical signals.
4. The sound quality is greatly affected by microphone technique. Microphone choice and placement are critical in this method.
5. The signals from the microphones go to a 2-track recorder. It may be a hard-drive recorder, CD-R burner, DVD-R burner, flash memory recorder, or

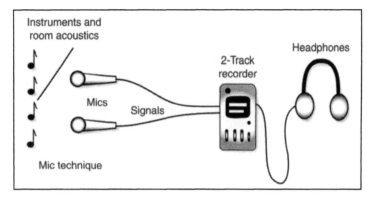

FIGURE 12.1
The recording chain for live stereo recording.

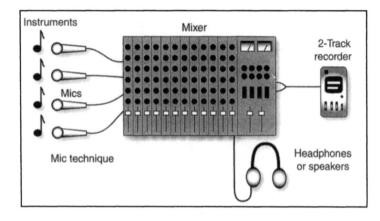

FIGURE 12.2
The recording chain for live-mix recording.

computer hard drive. The signal changes to a pattern stored on a medium, such as magnetic patterns on a hard disk. During playback, the patterns on the medium are converted back into a signal.

During recording, signals are stored along a track—a path or channel on the medium containing a recorded signal. One or more tracks can be recorded on a single medium. For example, a 2-track hard-disk recording stores two tracks on hard disk, such as the two different audio signals required for stereo recording.

6. To hear the signal being recorded, the engineer needs a monitor system: headphones or a stereo power amplifier and loudspeakers. Use the monitors to judge how well the mic technique is working.

The speakers or headphones convert the signal back into sound. This sound resembles that of the original instruments. Also, the acoustics of the listening room affect the sound reaching the listener.

Live-Mix Recording

This method is seldom used except for live broadcasts or recordings of PA mixes. Using a mixer, a mix of several microphones is set up and the engineer monitors and records the mixer's output signal. Each microphone is close to its sound source (Figure 12.2).

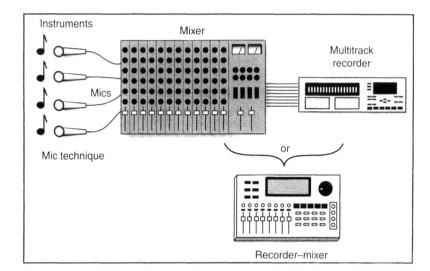

Instruments

Mixer

Multitrack recorder

Mics

Mic technique

or

Recorder–mixer

FIGURE 12.3
The recording chain for multitrack recording.

Multitrack Recorder and Mixer

The engineer records with several microphones into a mixer, which is connected to a multitrack recorder. The signal of each microphone is recorded on its own track. These recorded signals are then mixed after the performance is done. Different groups of instruments can be recorded on each track. Figure 12.3 shows the stages in this method:

1. Place microphones near the instruments.
2. Plug the microphones into a mixing console: a larger, perhaps more sophisticated mixer. During multitrack recording, the mixing console amplifies the weak microphone signals up to the level needed by the recorder. The console is also used to send each microphone signal to the desired track.
3. Record the amplified microphone signals on the multitrack recorder.

More instruments can be recorded later on unused tracks—a process called overdubbing. Wearing headphones, the performer listens to the recorded tracks and plays or sings along with them. This performance is then tracked to an unused track.

After the recording is done, the engineer will play all the tracks through the mixing console to mix them with a pleasing balance (Figure 12.4). Here are the steps:

1. Play back the multitrack recording of the song several times, adjusting the track volumes and tone controls until the mix is just right. One can add effects to enhance the sound quality. Some examples are echo, reverberation, and compression (discussed in Chapter 5). Effects are made by signal processors which connect to your mixer or by software applications that are part of a recording program.
2. Record or export the final mix on a 2-track stereo recorder (hard-disk recorder, memory recorder, CD-R, DVD-R, or computer hard drive).

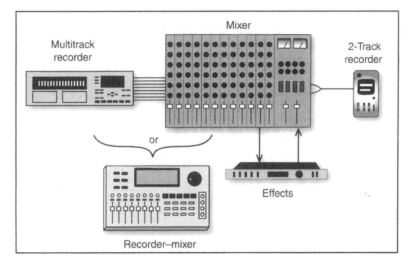

FIGURE 12.4
The recording chain for a multitrack mixdown.

FIGURE 12.5
TASCAM DP-02, an example of a recorder–mixer.

Two types of multitrack recorders are a hard-disk recorder and a flash memory recorder. They have random access: you can instantly go to any part of the recorded program (Figure 12.5).

Stand-Alone Digital Audio Workstation (Recorder–Mixer)

This is a multitrack recorder and a mixer combined in one portable chassis. It's relatively easy to use. The recorder is a hard drive or a flash memory card. Other names for a recorder–mixer are "stand-alone Digital Audio Workstation," "digital multitracker," or "portable studio." Most recorder–mixers have built-in effects.

Computer Digital Audio Workstation

This low-cost system includes a computer, recording software, and a sound card or audio interface that gets audio into and out of a computer (Figure 12.6). The computer's hard drive records the audio.

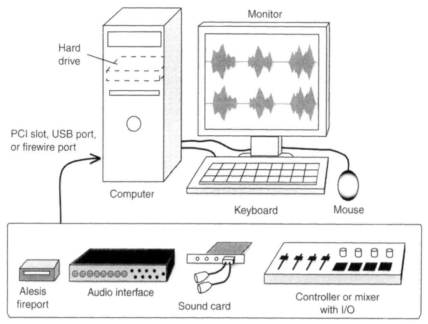

FIGURE 12.6
Computer with a choice of audio interface and recording–editing software.

The recording software allows these operations:

1. Record music to the computer's hard drive.
2. Edit the tracks to fix mistakes, delete unwanted material, or to copy/move song sections.
3. Mix the tracks with a mouse or controller by adjusting virtual controls that appear on your computer screen.

You might also assemble a song from samples or from loops. Samples are recordings of single notes of various instruments. Loops are repeating musical patterns.

MIDI Sequencing

With this recording method, a musician performs on a MIDI controller, such as a piano-style keyboard or drum pads. The controller puts out a MIDI signal, a series of numbers that indicates which keys were pressed and when they were pressed. The MIDI signal is recorded into computer memory by a sequencer or sequencer program in a computer. When you play back the MIDI sequence, it plays the tone generators in a synthesizer or sound module. A MIDI sequence can also play samples: digital recordings of musical notes played by real instruments.

Like a player piano, MIDI sequencing records your performance gestures rather than audio. Figure 12.7 shows the process.

MIDI/digital audio software offers the ability to record MIDI sequences and digital audio onto hard disk. The engineer first records a few tracks of MIDI onto hard disk, then adds audio tracks: lead vocal, sax solo, or whatever. All these elements stay synchronized.

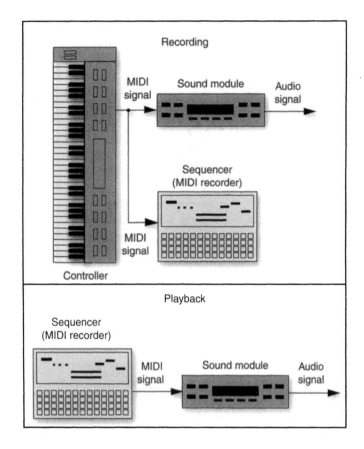

FIGURE 12.7
MIDI sequencing system.

PROS AND CONS OF EACH METHOD

Live stereo recording is simple, cheap, and fast. But it usually sounds too muddy with rock music, and you must adjust balances by moving musicians. It can work well with classical music, and sometimes with folk or acoustic jazz music.

Live-mix recording is fairly simple and quick. However, loud instruments might sound distant in the recording because their sound "leaks" into distant microphones. And if the mix or performance has mistakes, the band has to re-record the entire song. Also, the live sound of the band can make it hard to hear the monitored sound clearly.

Multitrack recording has many advantages. You can punch-in: fix a musical mistake by recording a new, correct part over the mistake. You can overdub: record one instrument at a time. This reduces leakage and gives a tighter sound. Also, you can postpone mixing decisions until after the performance. Then you can monitor the mix in quiet surroundings. This method is more complex and expensive than live-mix recording.

If you use a separate multitrack recorder and mixer, each component can be used independently. For example, an on-location live recording might be captured with just the recorder or a PA job with just the mixer. Or, if you already have a mixer, all you need to buy is a recorder. This system is a little difficult to set up because you need to connect cables between the mixer and the recorder, and between the mixer and the outboard effects units.

A stand-alone DAW (recorder–mixer) is easy to use because it is a single portable chassis that includes most of your studio equipment: recorder, mixer, effects, and often a CD burner. It doesn't require cables except for the microphones, instruments, and monitor speakers. High-end units let you edit the music. They also have automated mixing: memory chips in the mixer remember your mixdown settings, and reset the mixer accordingly the next time you play back the recording.

A computer DAW is inexpensive, powerful, and flexible. It lets you do sophisticated editing and automated mixing. Several plug-in (software) effects are included, and you can purchase and install other plug-ins. Recording software can be updated at little cost. As for drawbacks, computers can crash and can be difficult to set up and optimize for audio work.

MIDI sequencing lets you record musical parts by entering notes slowly or one-at-a-time if you wish. After the sequencing is finished, you can edit notes to correct mistakes. You can even change the instrument sounds or the tempo. However, you are limited to the sounds of samples and sound modules unless you use MIDI/digital audio software, which lets you add miked instruments to the mix.

RECORDING THE MIXES

No matter which recording method you use, eventually you'll mix each song and record the mix on a 2-track recorder, or record the mix as a stereo wave file on your hard drive. You can convert the wave file to an MP3 or WMA format for uploading to the Web.

You might want to assemble an album of your recorded mixes. You remove noises and count-offs between songs, put the songs in the desired order, and put a few seconds of silence between songs. This is done with a computer and editing software (a DAW, Figure 12.6). The last step is to copy the album to a blank CD. There's your final product, ready to duplicate.

No matter what type of recording you do, each stage contributes to the sound quality of the finished recording. A bad-sounding CD can be caused by any weak link: low-quality microphones, bad microphone placement, improperly set mixer controls, and so on. A great-sounding recording results when you get every stage right. This book will help you reach that goal.

FURTHER READING

This chapter is an edited extract from Bartlettt, B. & Bartlett, J. (2005). *Practical Recording Techniques*, Focal Press, Oxford.

Chapter 13
Evaluating Sound Quality

Bruce and Jenny Bartlett

In This Chapter

Our ability to critically listen is explored more in this chapter by Bruce and Jenny Bartlett.

INTRODUCTION

Later in the production process, you could seat an engineer behind a mixing console and ask him or her to do a mix. It might sound great. Then seat another engineer behind the same console and again ask for a mix. It sounds terrible. What happened?

The difference lies mainly in their ears—their critical listening ability. Some engineers have a clear idea of what they want to hear and how to get it. Others haven't acquired the essential ability to recognize good sound. By knowing what to listen for, you can improve your artistic judgments during recording and mixdown. You are able to hear errors in microphone placement, equalization, and so on, and correct them.

To train your hearing, try to analyze recorded sound into its components—such as frequency response, noise, reverberation—and concentrate on each one in turn. It's easier to hear sonic flaws if you focus on a single aspect of sound reproduction at a time. This chapter is a guide to help you do this.

STEREO RECORDING VS. MULTITRACK RECORDING

Acoustic-based music, such as classical, and popular music have different standards of "good sound." One goal in recording classical music (and sometimes other acoustic-based music such as folk or some forms of jazz) is to accurately reproduce the live performance. This is a worthy aim because the sound of an orchestra in a good hall can be quite beautiful. The music was composed and the instruments were designed to sound best when heard live in a concert hall. The recording engineer, out of respect for the music, should always try to translate that sound to disc with as little technical intrusion as possible.

By contrast, the accurate translation of sound to disc is not always the goal in recording popular music. Although the aim may be to reproduce the original sound, the producer or engineer may also want to play with that sound to create a new sonic experience or to do some of both.

In fact, the artistic manipulation of sounds through studio techniques has become an end in itself. Creating an interesting new sound is as valid a goal as re-creating the original sound. There are two games to play, each with its own measures of success.

If the aim of a recording is realism or accurate reproduction, the recording is successful when it matches the live performance heard in the best seat in the concert hall. The sound of musical instruments is the standard by which such recordings are judged.

When the goal is to enhance the sound or produce special effects (as in most pop-music recordings), the desired sonic effect is less defined. The live sound of a pop group could be a reference, but pop-music recordings generally sound better than live performances—recorded vocals are clearer and less harsh, the bass is cleaner and tighter, and so on. The sound of pop music reproduced over speakers has developed its own standards of quality apart from accurate reproduction.

GOOD SOUND IN A POP-MUSIC RECORDING

Currently, a good-sounding pop recording might be described in the following terms (there are always exceptions). These can provide an excellent dictionary of terms with which to communicate with your production team:

- Well mixed
- Wide range
- Tonally balanced
- Clean

- Clear
- Smooth
- Spacious.

It also has:

- Presence
- Sharp transients
- Tight bass and drums
- Wide and detailed stereo imaging
- Wide but controlled dynamic range
- Interesting sounds
- Suitable production.

The following sections explore each one of these qualities in detail so that you know what to listen for. Assume that the monitor system is accurate, so that any colorations heard are in the recording and not in the monitors.

A Good Mix

In a good mix, the loudness of instruments and vocals is in a pleasing balance. Everything can be clearly heard, yet nothing is obtrusive. The most important instruments or voices are loudest; less important parts are in the background.

A successful mix goes unnoticed. When all the tracks are balanced correctly, nothing sticks out and nothing is hidden. Of course, there's a wide latitude for musical interpretation and personal taste in making a mix. Dance mixes, for example, can be very severe sonically.

Sometimes you don't want everything to be clearly heard. On rare occasions you may want to mix in certain tracks very subtly for a subconscious effect.

The mix must be appropriate for the style of music. For example, a mix that's right for loud rock music usually won't work for a pop ballad. A rock mix typically has the drums way up front and the vocals only slightly louder than the accompaniment. In contrast, a pop ballad has the vocals loudest, with the drums used just as "seasoning" in the background.

Level changes during the mix should be subtle or should make sense. Otherwise, instruments jump out for a solo and fall back in afterward. Move faders slowly or set them to preset positions during pauses in the music. Nothing sounds more amateurish than a solo that starts too quietly and then comes up as it plays—you can hear the engineer working the fader.

Wide Range

Wide range means extended low- and high-frequency response. Cymbals should sound crisp and distinct, but not sizzly or harsh; kick drum and bass should sound deep, but not overwhelming or muddy. Wide-range sound results from using high-quality microphones and adequate EQ.

You might want to combine "hi-fi" and "low-fi" sounds in a single mix. The low-fi sounds generally cover a narrow frequency range and might be distorted or noisy.

Good Tonal Balance

The overall tonal balance of a recording should be neither bassy nor trebly. That is, the perceived spectrum should not emphasize low or high frequencies. Low bass, midbass, midrange, upper midrange, and highs should be heard in equal proportions (Figure 13.1). Emphasis of any one frequency band over the other eventually causes listening fatigue. Dance club mixes, however, are heavy on the bass end to get the crowd moving.

Recorded tonal balance is inversely related to the frequency response of the studio's monitor system. If the monitors have a high-frequency roll-off, the engineer will compensate by boosting highs in the recording to make the monitors sound correct. The result is a bright recording.

Before doing a mix, play over the monitors some commercial recordings whose sound you admire, to become accustomed to a commercial spectral balance. After your mix is recorded, play it back and alternately switch between your mix and a commercial recording. This comparison indicates how well you matched a commercial spectral balance. Of course, you may not care to duplicate what others are doing. To compare your spectral balance to others and improve it, you might try Harmonic Balancer (www.har-bal.com).

In pop-music recordings, the tonal balance or timbre of instruments does not necessarily have to be natural. Still, many listeners want to hear a realistic timbre from acoustic instruments, such as the guitar, flute, sax, or piano. The reproduced timbre depends on microphone frequency response, microphone placement, the musical instruments themselves, and equalization.

Clean Sound

Clean means free of noise and distortion. Hiss, hum, and distortion are inaudible in a good recording. Distortion in this case means distortion added by the recording

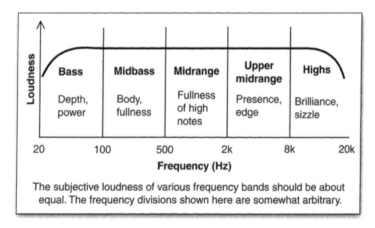

FIGURE 13.1
Loudness vs. frequency of a pop recording with good sound.

process, not distortion already present in the sound of electric-guitar amps or Leslie speakers. There are exceptions to this guideline; some popular recordings have noise or distortion added intentionally.

Clean also means "not muddy" or free of low-frequency ringing and leakage. A clean mix is one that is uncluttered or free of excess instrumentation. This is achieved by arranging the music so that similar parts don't overlap, and not too many instruments play at once in the same frequency range. Usually, the fewer the instruments, the cleaner the sound. Too many overdubs can muddy the mix.

Clarity

In a clear-sounding recording, instruments do not crowd or mask each other. They are separate and distinct. As with a clean sound, clarity arises when instrumentation is sparse, or when instruments occupy different areas of the frequency spectrum. For example, the bass provides low frequencies, keyboards might emphasize midbass, lead guitar provides upper midrange, and cymbals fill in the highs.

In addition, a clear recording has adequate reproduction of each instrument's harmonics. That is, the high-frequency response is not rolled off.

Smoothness

Smooth means easy on the ears, not harsh, uncolored. Sibilant sounds are clear but not piercing. A smooth, effortless sound allows relaxation; a strained or irritating sound causes muscle tension in the ears or body. Smoothness is a lack of sharp peaks or dips in the frequency response, as well as a lack of excessive boost in the midrange or upper midrange. It is also low distortion, such as provided by a 24-bit recording.

Presence

Presence is the apparent sense of closeness of the instruments—a feeling that they are present in the listening room. Synonyms are clarity, detail, and punch.

Presence is achieved by close miking, overdubbing, and using microphones with a presence peak or emphasis around 5 kHz. Using less reverb and effects can help. Upper-midrange boost helps, too. Most instruments have a frequency range that, if boosted, makes the instrument stand out more clearly or become better defined. Presence sometimes conflicts with smoothness because presence often involves an upper-midrange boost, while a smooth sound is free of such emphasis. You have to find a tasteful compromise between the two.

Spaciousness

When the sound is spacious or airy, there is a sense of air around the instruments. Without air or ambience, instruments sound as if they are isolated in stuffed closets. (Sometimes, though, this is the desired effect.) You achieve spaciousness by adding reverb, recording instruments in stereo, using room mics, or miking farther away.

Sharp Transients

The attack of cymbals and drums generally should be sharp and clear. A bass guitar and piano may or may not require sharp attacks, depending on the song.

Tight Bass and Drums

The kick drum and bass guitar should "lock" together so that they sound like a single instrument—a bass with a percussive attack. The drummer and bassist should work out their parts together so they hit accents simultaneously, if this is desired.

To further tighten the sound, damp the kick drum and record the bass direct. Rap music, however, has its own sound—the kick drum usually is undamped and boomy, sometimes with short reverb added. Equalize the kick and bass in complementary ways so that they don't mask each other, for example:

Kick: Boost 60 to 80 Hz, cut 150 to 400 Hz, boost 3 kHz.

Bass: Cut 60 to 80 Hz, boost 120 to 150 Hz, boost 900 Hz.

Wide and Detailed Stereo Imaging

Stereo means more than just left and right. Usually, tracks should be panned to many points across the stereo stage between the monitor speakers. Some instruments should be hard left or hard right, some should be in the center, others should be half-left or half-right. Try to achieve a stereo stage that is well balanced between left and right (Figure 13.2). Instruments that occupy the same frequency range can be made more distinct by panning them to opposite sides of center.

You may want some tracks to be unlocalized. Backup choruses and strings should be spread out rather than appearing as point sources. Stereo keyboard sounds can wander between speakers. A lead guitar solo can have a fat, spacious sound.

There should also be some front-to-back depth. Some instruments should sound close or up front; others should sound farther away. Use different miking distances or different amounts of reverb on various tracks.

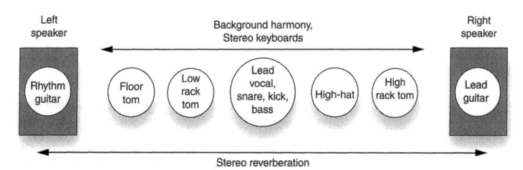

FIGURE 13.2
An example of image placement between speakers.

©2009 Bruce and Jenny Bartlett

If you want the stereo imaging to be realistic (for a jazz combo, for example), the reproduced ensemble should simulate the spatial layout of the live ensemble. If you're sitting in an audience listening to a jazz quartet, you might hear drums on the left, piano on the right, bass in the middle, and sax slightly right. The drums and piano are not point sources, but are somewhat spread out. If spatial realism is the goal, you should hear the same ensemble layout between your speakers. On some commercial CDs, the piano and drums are spread all the way between speakers— an interesting effect, but unrealistic.

Pan-potted mono tracks often sound artificial in that each instrument sounds isolated in its own little space. It helps to add some stereo reverberation around the instruments to "glue" them together.

Often, TV mixes are heard in mono. Hard-panned signals sound weak in mono relative to center-panned signals. So pan sound sources to 3 and 9 o'clock, not hard right and hard left if you desire true mono compatibility. Check your music in mono to see if your wide stereo parts are being compromised.

Large but Controlled Dynamic Range

Dynamic range is the range of volume levels from softest to loudest. A recording with a large dynamic range becomes noticeably louder and softer, adding excitement to the music. To achieve this, one shouldn't add too much compression (which acts like an automatic volume control). Overly compressed signals sound squashed so crescendos and quiet interludes lose their impact, and the sound becomes fatiguing.

Vocals often need some compression or gain-riding because they have more dynamic range than the instrumental backup. A vocalist may sing too loudly and blast the listener, or sing too softly and become buried in the mix. A compressor can even out these extreme level variations, keeping the vocals at a constant loudness. Bass guitar also can benefit from compression.

Interesting Sounds

The recorded sound may be too flat or neutral, lacking character or color. In contrast, a recording with creative production has unique musical instrument sounds, and typically uses effects. Some of these are equalization, echo, reverberation, doubling, chorus, flanging, compression, distortion, and stereo effects.

Making sounds interesting or colorful can conflict with accuracy or fidelity. That's okay, but you should know the trade-off and prepare accordingly.

Suitable Production

The way a recording sounds should imply the same message as the musical style or lyrics. In other words, the sound should be appropriate for the particular tune being recorded.

For example, some rock music is rough and raw. The sound should be, too. A clean, polished production doesn't always work for high-energy rock 'n' roll. There might even be a lot of leakage or ambience to suggest a garage studio or nightclub environment. The role of the drums is important, so they should be loud in the mix. The toms should ring, if that is desired.

New Age, disco, R&B, contemporary Christian, or pop music is slickly produced. The sound is usually tight, smooth, and spacious. Folk music and acoustic jazz typically sound natural and have little or no reverb. Music in the digital hardcore genre has lots of distortion.

Actually, each style of music is not locked into a particular style of production. You tailor the sound to complement the music of each individual tune. Doing this may break some of the guidelines of good sound, but that's usually okay as long as the song is enhanced by its sonic presentation.

GOOD SOUND IN A CLASSICAL RECORDING

As with pop music, classical should sound clean, wide-range, and tonally balanced. Classical music recordings, for example, are meant to sound realistic—like a live performance. Therefore, they require good acoustics, a natural balance, tonal accuracy, suitable perspective, and accurate stereo imaging.

Good Acoustics

The acoustics of the concert hall or recital hall, or any other acoustic recording environment, should be appropriate for the style of music to be performed. Specifically, the reverberation time should be neither too short (dry) nor too long (cavernous). Too short a reverberation time results in a recording without spaciousness or grandeur. Too long a reverberation time blurs notes together, giving a muddy, washed-out effect. Ideal reverberation times are around 1.2 seconds for chamber music or soloists, 1.5 seconds for symphonic works, and 2 seconds for organ recitals. To get a rough idea of the reverb time of a room, clap your hands once, loudly, and count the seconds it takes for the reverb to fade to silence. For more information on this, which is known as the reverberation time, or T60, please refer to the further reading list for recording techniques.

A Natural Balance

When an acoustic recording is well balanced, the relative loudness of instruments is similar to that heard in an ideal seat in the audience area. For example, violins are not too loud or soft compared to the rest of the orchestra; harmonizing or contrapuntal melody lines are in proportion.

Generally, the conductor, composer, and musicians balance the music acoustically, and you capture that balance with your stereo mic pair. But sometimes you need to mic certain instruments or sections close-up with "spot mics" to enhance definition or balance. Then you mix all the mics. In either case, consult the conductor for proper balances.

Tonal Accuracy

The reproduced timbre or tone quality should match that of live instruments. Fundamentals and harmonics should be reproduced in their original proportion for classical material. With other acoustic performances, artistic license might be permitted for some change as appropriate.

Suitable Perspective

Perspective is the sense of distance of the performers from the listener. In other words how far away the stage sounds. Do the performers sound like they're eight rows in front of you, in your lap, in another room?

The style of music suggests a suitable perspective. Incisive, rhythmically motivated works (such as Stravinsky's *Rite of Spring*) sound best with closer miking; lush, romantic pieces (a Bruckner symphony) are best served by more distant miking. The chosen perspective depends on the taste of the producer.

Closely related to perspective is the amount of recorded ambience or reverberation. A good miking distance yields a pleasing balance of direct sound from the orchestra and ambience from the concert hall.

Accurate Imaging

Reproduced instruments should appear in the same relative locations as they were in the live performance. Instruments in the center of the ensemble should be heard in the center between the speakers; instruments at the left or right side of the ensemble should be heard from the left or right speaker. Instruments halfway to one side should be heard halfway off center, and so on. A large ensemble should spread from speaker to speaker, while a quartet or soloist can have a narrower spread.

It's important to sit equidistant from the speakers when judging stereo imaging, otherwise the images shift toward the side on which you're sitting. Sit as far from the speakers as they are spaced apart. Then the speakers appear to be 60 degrees apart, which is about the same angle an orchestra fills when viewed from the typical ideal seat in the audience (tenth row center, for example).

The reproduced size of an acoustic instrument or instrumental section should match its size in real life. A guitar should be a point source; a piano or string section should have some stereo spread. Each instrument's location should be as clearly defined as it was heard from the ideal seat in the concert hall.

Reproduced reverberation (concert-hall ambience) should surround the listener, or at least it should spread evenly between the speakers. Surround sound technology is needed to make the recorded ambience actually surround the listener, although spaced microphone recordings have some of this effect. Accurate imaging is illustrated in Figure 13.3.

There should be a sense of stage depth, with front-row instruments sounding closer than back-row instruments.

TRAINING YOUR HEARING

The critical process is easier if you focus on one aspect of sound reproduction at a time, as is discussed in Chapter 2. You might concentrate first on the tonal

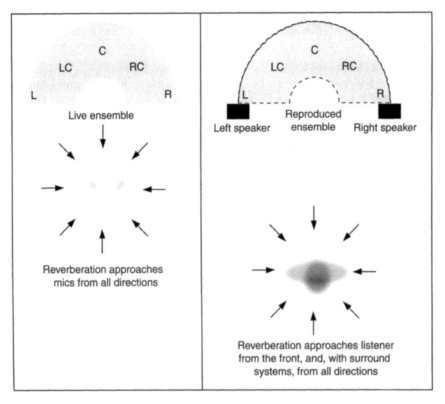

FIGURE 13.3
With accurate imaging, the sound-source location and size, and the reverberant field, are reproduced during playback.

balance—try to pinpoint what frequency ranges are being emphasized or slighted. Next listen to the mix, the clarity, and so on. Soon you have a lengthy description of the sound quality of your recording.

Developing an analytical ear is a continuing learning process. Train your hearing by listening carefully to recordings—both good and bad. Make a checklist of all the qualities mentioned in this chapter. Compare your own recordings to live instruments and to commercial recordings. Check out the Golden-Ear training CDs at www.moultonlabs.com/gold.htm.

A pop-music record that excels in all the attributes of good sound is The Sheffield Track Record (Sheffield Labs, Lab 20), engineered and produced by Bill Schnee. In effect, it's a course in state-of-the-art sound—required listening for any recording engineer or producer.

Another record with brilliant production is The Nightfly by Donald Fagen (Warner Brothers 23696-2)—engineered by Roger Nichols, Daniel Lazerus, and Elliot Scheiner; produced by Gary Katz; and mastered by Bob Ludwig. This recording, and Steely Dan recordings by Roger Nichols, sound razor sharp, very tight and clear, elegant, and tasteful; and the music just pops out of the speakers.

The following listings are more examples of outstanding rock production, and set high standards:

Songs:

- "I Need Somebody," by Bryan Adams; producer, Bob Clearmountain
- "The Power of Love," by Huey Lewis & The News; producer, Huey Lewis & The News.

Albums:

- 90125, by Yes; producer, Trevor Horn
- Synchronicity, by The Police; engineer and producer, Hugh Padgham and The Police
- Dark Side of the Moon, by Pink Floyd; recording engineer & producer, Alan Parsons
- Thriller, by Michael Jackson; engineer, Bruce Swedien; producer, Quincy Jones
- Avalon, by Roxy Music; engineer, Bob Clearmountain; producer, Roxy Music
- Nevermind, by Nirvana; producer, Butch Vig
- Come Away With Me, by Norah Jones; engineer, Jay Newland
- Genius Loves Company, by Ray Charles; several engineers
- Give, by The Bad Plus; engineer, Tchad Blake
- Live in Paris and The Look of Love, by Diana Krall; engineer, Al Schmitt
- Smile, by Brian Wilson; engineer, Mark Linett.

Some other recordings of note are those by the groups Supertramp, Heart and Radiohead; and by producers George Martin and Andy Wallace.

Then there are the incredibly clean recordings of Tom Jung (with DMP records) and George Massenberg. Some classical music recordings with outstanding sound are on the Telarc, Delos, and Chesky labels. You can learn a lot by emulating these superb recordings and many others.

Once you're making recordings that are competent technically—clean, natural, and well mixed—the next stage is to produce imaginative sounds. You're in command; you can tailor the mix, through your production team, to sound any way that pleases you or your band during the process. The supreme achievement is to produce a recording that is a sonic knockout, beautiful, or thrilling.

TROUBLESHOOTING BAD SOUND

Now you know how to recognize good sound, but can you recognize bad sound? Suppose you're listening to a recording in progress, or listening to a recording you've already made. Something doesn't sound right. How can you pinpoint what's wrong, and how can you get it fixed?

The rest of this chapter is intended as a guide for you to "hear" what might be going wrong. With this knowledge on board, you can understand the finer workings of the recording process and discuss more freely with your production team, the problems you're hearing.

This troubleshooting guide is divided into four main sections:

1. Bad sound on all recordings
2. Bad sound on playback only (the mixer output sounds all right)
3. Bad sound in a pop-music recording
4. Bad sound in classical music recording.

Before you get to the studio you should check your equipment for faulty cables and connectors. Also check all control positions; rotate knobs, and flip switches to clean the contacts, and clean connectors with DeoxIT from Caig Labs. All these things will save precious (and expensive) studio time down the line.

Bad Sound on All Recordings in the Studio

If you note your recordings are not turning out as they should in the session, start to consider these issues:

- Is the monitor system suitable for the material being recorded? (Auratone speakers, a very small near field monitor, are not necessarily the best monitors for drum 'n' bass, for example.)
- If so suggest using the main monitors.
- Improve room acoustics. It is presumed that the studio has been acoustically designed and does not need or deserve this comment. However, some studios may still require some attention and here's some hints:
  □ Move the speakers
  □ Equalize the monitor system
  □ Try different models of monitor speakers
  □ Upgrade the power amp and speaker cables
  □ Monitor at a moderate listening volume, such as 85 dB SPL. We hear less bass and treble in a program if it is monitored at a low volume, and vice versa. If we hear too little bass due to monitoring at a low level, we might mix in too much bass.

Bad Sound on Playback Only

You get the mixes home and they sound lackluster. You might have bad sound on your playback only—the studio output sounded okay. If Digital Audio Tape (DAT) playback has glitches or drop-outs, try these steps:

- Clean the recorder with a dry cleaning tape.
- Before recording, fast-forward the tape to the end and rewind it to the top.
- Use better tape.
- If connected digitally, check for digital clocking issues. Does the DAT machine need to be the wordclock master?

If a hard drive-based system is used for recording, it can present glitches or drop-outs on playback. If this happens it is likely that the buffer size of the system might need to be increased. This can usually be found in the preference panes of the software.

If a digital recording sounds distorted, these suggestions might be a potential solution:

- The recording level should be as high as possible, but should not exceed $-6\,dBFS$ (decibels Full Scale) in peak meter mode, not rms mode. This is not the same as in analog recording where the meters could comfortably be in the red on occasion. In digital audio, the level should never hit the red.
- Avoid clipping in effects plug-ins or along the signal path.
- Ask to record at a higher sampling rate or higher bit depth.
- Avoid sample-rate conversions where possible until the very end of the signal chain at mastering say, and apply dither when going from a high bit depth to a lower bit depth.
- Try an alternative higher quality audio interface.
- Ask whether the wordclock is synchronized properly if you're using a digital console and Digital Audio Workstation (DAW).

Bad Sound in a Pop-Music Recording Session

Sometimes you have bad sound in a pop-music recording session.

MUDDINESS (LEAKAGE)

If the sound is muddy from excessive leakage, try the following:

- Perhaps place the microphones closer to their sound sources.
- Spread the instruments farther apart to reduce the level of the leakage.
- Place the instruments closer together to reduce the delay of the leakage.
- Use directional microphones (such as cardioids).
- Overdub the instruments.
- Record the electric instruments direct.
- Use baffles (gobos) between instruments.
- Deaden the room acoustics (add absorptive material or flexible panels to the walls, such as bass traps, and other acoustic treatments).
- Filter out frequencies above and below the spectral range of each instrument. Be careful or you'll change the sound of the instrument.
- Try things such as turning down the bass amp in the studio, use a DI Box.

MUDDINESS (EXCESSIVE REVERBERATION)

If the sound is muddy due to excessive reverberation, try these steps:

- Reduce the use of effects, or lower the effects send levels. Alternatively don't use effects until you figure out what the real problem is.
- Place the microphones closer to their sound sources.
- Use directional microphones (such as cardioids).
- Deaden the room acoustics.
- Filter out frequencies below the fundamental frequency of each instrument; in theory these are not part of the sound of this instrument. (It is worth noting that some instruments such as the Hammond Organ and percussion do have frequencies that are sometimes lower than the intended fundamental.)

MUDDINESS (LACKS HIGHS)

If your sound is muddy and lacks highs, or has a dull or muffled sound, try the following:

- Use microphones with better high-frequency response, or use condenser mics instead of dynamics.
- Change the microphone placement. Put the mic in a spot where there are sufficient high frequencies. Keep the high-frequency sources (such as cymbals) on-axis to the microphones.
- Use small diameter microphones, which generally have a flatter response off-axis.
- Boost the high-frequency equalization or cut the lower frequencies slightly around 300 Hz.
- Change musical instruments; replace guitar strings; replace drum heads.
- Ask to use an enhancer signal processor, but watch out for noise.
- Use a direct box on the electric bass. Have the bassist play percussively or use a pick if the music requires it. When compressing the bass, use a long attack time to allow the note's attack to come through. (Some songs don't require sharp bass attacks—do whatever's right for the song.)
- Damp the kick drum with a pillow, folded towel or blanket, and mike it next to the center of the head near the beater. Use a wooden or plastic beater if the song and the drummer allow it.
- Don't plug an electric guitar directly into a microphone input. Use a direct box or a high-impedance input.
- Apply high-frequency boost after compression, not before.

MUDDINESS (LACKS CLARITY)

If your sound is muddy because it lacks clarity, try these steps:

- Consider using fewer instruments in the musical arrangement. Maybe turn down synth pads in the mix.
- Equalize instruments differently so that their spectra don't overlap.
- Try less reverberation.
- Using equalizers, boost the presence range of instruments that lack clarity. Or cut 1 to 2 dB around 300 Hz.
- In a reverb unit, add about 45 to 100 milliseconds of pre-delay.
- Pan similar sounding instruments to opposite sides.

DISTORTION

If you hear distortion when monitoring the microphones in a pop-music recording, try the following ideas:

- Increase input attenuation (reduce input gain), or plug in a pad between the microphone and microphone input.
- Readjust gain-staging: Set faders and pots to their design centers (shaded areas).
- If you still hear distortion, switch in the pad built into the microphone (if any).

- Check connectors for stray wires and bad solder joints.
- Unplug and plug in connectors. Clean them with Caig Labs DeoxIT or Pro Gold.

TONAL IMBALANCE

If you have bad tonal balance—the sound is boomy, dull, or shrill, for example—try these steps: change musical instruments, guitar strings, reeds, and so on.

- Change microphone placement. If the sound is too bassy with a directional microphone, you may be getting proximity effect. Mike farther away or roll off the excess bass.
- Use the 3:1 rule of microphone placement to avoid phase cancellations. When you mix two or more microphones to the same channel, the distance between microphones should be at least 3 times the microphone-to-source distance.
- Try another microphone. If the proximity effect of a cardioid microphone is causing a bass boost, try an omnidirectional microphone instead.
- If you must place a microphone near a hard, reflective surface, try a boundary microphone on the surface to prevent phase cancellations.
- If you're recording a singer/guitarist, delay the vocal microphone signal by about 1 millisecond.
- Change the equalization. Avoid excessive boost. Maybe cut slightly around 300 Hz if the sound is muddy or cut around 3 kHz if the sound is harsh.
- Use equalizers with a broad bandwidth, rather than a narrow, peaked response.

LIFELESSNESS

If your pop-music recording has a lifeless sound and is unexciting, these steps might help you solve it:

- Work on the live sound of the instruments in the studio to come up with unique effects.
- Add effects: reverberation, echo, exciter, doubling, equalization, and so on.
- Use and combine recording equipment in unusual ways.
- Try overdubbing little vocal licks or synthesized sound effects.

If your sound seems lifeless due to dry or dead acoustics, try these:

- If leakage is not a problem, put microphones far enough from instruments to pick up wall reflections. If you don't like the sound this produces, try the next suggestion.
- Add reverb or echo to dry tracks. (Not all tracks require reverberation. Also, some songs may need very little reverberation so that they sound intimate.)
- Use omnidirectional microphones to capture more ambience.
- Add hard, reflective surfaces in the studio, or record in a hard-walled room.
- Allow a little leakage between microphones. Put microphones far enough from instruments to pick up off-microphone sounds from other instruments. This should not be overdone or the sound becomes muddy and track separation becomes poor.

NOISE (HISS)

Sometimes your pop-music recording has extra noise on it. If your sound has hiss, try these:

- Check for noisy guitar amps or keyboards.
- Switch out the pad built into the microphone (if any).
- Reduce mixer input attenuation (increase input gain).
- Use a more sensitive microphone.
- Use an impedance-matching adapter (a low- to high-Z step-up transformer) between microphones and phone-jack microphone inputs.
- Use a quieter microphone (one with low self-noise).
- Increase the sound pressure level at the microphone by miking closer. If you're using PZMs, mount them on a large surface or in a corner.
- If possible, feed recorder tracks from mixer direct outs or insert sends instead of group or bus outputs.
- Use a low-pass filter (high-cut filter).
- As a last resort, use a noise gate.

NOISE (RUMBLE)

If the noise is a low-frequency rumble, follow these steps:

- Reduce air-conditioning noise or shut off the air conditioning temporarily.
- Use a high-pass filter (low-cut filter) that is set around 40–80 Hz.
- Use microphones with limited low-frequency response.
- See the section Noise (Thumps).

NOISE (THUMPS)

- Change the microphone position.
- Change the musical instrument.
- Use a high-pass filter set around 40–80 Hz.
- If the cause is mechanical vibration traveling up the microphone stand, put the microphone in a shock-mount stand adapter or place the microphone stand on some carpet padding. Try to use a microphone that is less susceptible to mechanical vibration, such as an omnidirectional microphone or a unidirectional microphone with a good internal shock mount.
- Use a microphone with a limited low-frequency response.
- If the cause is piano pedal thumps, also try working on the pedal mechanism.

HUM

- To prevent ground loops, plug all equipment into outlet strips powered by the same AC outlet. Make sure that the sum of the equipment current ratings does not exceed the breaker's amp rating for that outlet.
- Some power amps create hum if they don't get enough AC current. So connect the power amp (or powered speakers) AC plug to its own wall outlet socket— the same outlet that feeds the outlet strips for the recording equipment.

- If possible, use balanced cables going into balanced equipment. Balanced cables have XLR or TRS connectors and two conductors surrounded by a shield. Ideally, the shield should be connected to the chassis ground (not the signal ground) at both ends of the cable.
- Transformer-isolate unbalanced connections.
- Don't use conventional SCR dimmers to change the studio lighting levels. Use Luxtrol® variable-transformer dimmers or multi-way incandescent bulbs instead.

Even if your system is wired properly, a hum or buzz may appear when you make a connection. Follow these tips to stop the hum:

- If the hum is coming from a direct box, flip its ground-lift switch.
- Check cables and connectors for broken leads and shields.
- Unplug all equipment from each other. Start by listening just to the powered monitor speakers. Connect a component to the system one at a time and see when the hum starts.
- Remove audio cables from your devices and monitor each device by itself. It may be defective.
- Lower the volume on your power amp (or powered speakers) and feed them a higher level signal.
- Use a direct box instead of a guitar cord between instrument and mixer.
- To stop a ground loop when connecting two devices, connect between them a 1:1 isolation transformer, direct box or hum eliminator (such as Jensen or Ebtech).
- Make sure that the snake box is not touching metal.
- To prevent accidental ground loops, do not connect XLR pin 1 to the connector shell except for permanent connections to equipment inputs and outputs.
- Try another microphone.
- If you hear a hum or buzz from an electric guitar, have the player move to a different location or aim in a different direction. You might also attach a wire between the player's body and the guitar strings near the tailpiece to ground the player's body.
- Turn down the high-frequency EQ on a buzzing bass guitar track.
- To reduce buzzing between notes on an electric guitar track, apply a noise gate.
- Route microphone cables and patch cords away from power cords; separate them vertically where they cross. Also keep recording equipment and cables away from computer monitors, power amplifiers, and power transformers.
- See Rane's excellent article on sound system interconnections at www.rane.com.

POPS

Pops are explosive breath sounds in a vocalist's microphone. If your pop-music recording has pops, try these solutions:

- Place the microphone above or to the side of the mouth.
- Place a foam windscreen (pop filter) on the microphone.
- Use a Pop Shield: Stretch a nylon stocking over a crochet hoop, and mount it on a microphone stand a few inches from the microphone (or use an equivalent commercial product).

- Place the microphone farther from the vocalist.
- Use a microphone with a built-in pop filter (ball grille).
- Use an omnidirectional microphone, because it is likely to pop less than a directional (cardioid) microphone.
- Switch in a high-pass filter (low-cut filter) set around 80 Hz.

SIBILANCE

Sibilance is an overemphasis of "s" and "sh" sounds. If you are getting sibilance on your pop music recording, try these steps:

- Use a de-esser signal processor or plug-in. Or use a multiband compressor, and compress the range 5–10 kHz.
- Place the microphone farther from the vocalist.
- Place the microphone toward one side of the vocalist, or at nose height, rather than directly in front.
- Cut equalization in the range 5–10 kHz.
- Change to a duller-sounding microphone such as a ribbon.

BAD MIX

Some instruments or voices are too loud or too quiet. To improve a bad mix, try the following:

- Change the mix. (Maybe change the mix engineer!)
- Compress vocals or instruments that occasionally get buried.
- Change the equalization on certain instruments to help them stand out.
- During mixdown, continuously change the mix to highlight certain instruments according to the demands of the music.
- Change the musical arrangement so that different musical parts don't play at the same time. That is, consider having a call-and-response arrangement (fill-in-the-holes) instead of everything playing at once, all the time.

UNNATURAL DYNAMICS

When your pop-music recording has unnatural dynamics, loud sounds don't get loud enough. If this happens, try these steps:

- Use less compression or limiting.
- Avoid overall compression.
- If buss compression must be used: try using multiband compression on the stereo mix instead of wideband (full-range) compression.

ISOLATED SOUND

If some of the instruments on your recording sound too isolated, as if they are not in the same room as the others, follow these steps:

- In general, allow a little crosstalk between the left and right channels. If tracks are totally isolated, it's hard to achieve the illusion that all the instruments are playing in the same room at the same time. You need some crosstalk or correlation between channels. Some right-channel information should leak into the left channel, and vice versa.

- Place microphones farther from their sound sources to increase leakage.
- Use omnidirectional microphones to increase leakage.
- Use stereo reverberation or echo.
- Pan effects return to the channel opposite the channel of the dry sound source.
- Pan extreme left and right tracks slightly toward center.
- Make the effects send levels more similar for various tracks.
- To give a lead guitar solo a fat, spacious sound, use a stereo chorus. Or send its signal through a delay unit, pan the direct sound hard left, and pan the delayed sound hard right.

LACK OF DEPTH

If the mix lacks depth, try these steps:

- Achieve depth by miking instruments at different distances.
- Use varied amounts of reverberation on each instrument. The higher the ratio of reverberant sound to direct sound, the more distant the track sounds.
- To make instruments sound closer, use a longer pre-delay setting in your reverb (maybe 40–100 milliseconds). Use a shorter pre-delay (under 30 milliseconds) to make instruments sound farther away.

Bad Sound in a Classical Music Recording

Check the following procedures if you have problems recording classical acoustic-based music.

TOO DEAD

If the sound in your acoustic recording is too dead—there is not enough ambience or reverberation—try these measures to solve the problem:

- Place the microphones farther from the performers.
- Use omnidirectional microphones.
- Record in a concert hall with better acoustics (longer reverberation time).
- Turn up the hall microphones (if used).
- Add artificial reverberation.

TOO CLOSE

If the sound is too detailed, too close, or too edgy, follow these steps:

- Place the microphones farther from the performers.
- Place the microphones lower or on the floor (as with a boundary microphone).
- Roll off the high frequencies.
- Use mellow-sounding microphones (many ribbon microphones have this quality).
- Turn up the hall microphones (if used).
- Increase the reverb send level.

©2009 Bruce and Jenny Bartlett

TOO DISTANT

If the sound is distant and there is too much reverberation, these steps might help:

- Place the microphones closer to the performers.
- Use directional microphones (such as cardioids).
- Record in a concert hall that is less live (reverberant).
- Turn down the hall microphones (if used).
- Decrease the reverb send level.

STEREO SPREAD TOO NARROW OR TOO WIDE

If your recording has a narrow stereo spread, try these steps:

- Angle or space the main microphone pair farther apart.
- If you're doing mid-side (MS) stereo recording, turn up the side output of the stereo microphone setup. For more information on MS setups, please consult the further reading list.
- Place the main microphone pair closer to the ensemble.

If the sound has excessive stereo spread (or "hole-in-the-middle"), try the following:

- Angle or space the main microphone pair closer together.
- If you're doing mid-side stereo recording, turn down the side output of the stereo microphone.
- In spaced-pair recording, add a microphone midway between the outer pair and pan its signal to the center.
- Place the microphones farther from the performers.

LACK OF DEPTH

Try the following to bring more depth into your acoustic-based music recording:

- Use only a single pair of microphones out front. Avoid multimiking.
- If you must use spot microphones, keep their level low in the mix.
- Add more artificial reverberation to the distant instruments than to the close instruments.

BAD BALANCE

If your recording has bad balance, try the following:

- Place the microphones higher or farther from the performer. Ask the conductor or performers to change the instruments' written dynamics. Be tactful!
- Add spot microphones close to instruments or sections needing reinforcement. Mix them in subtly with the main microphones' signals.

MUDDY BASS

If your recording has a muddy bass sound, follow these steps:

- Aim the bass-drum head at the microphones.
- Put the microphone stands and bass-drum stand on resilient isolation mounts (such as a carpet pad), or place the microphones in shock-mount stand adapters.
- Roll off the low frequencies or use a high-pass filter set around 40–80 Hz.
- Use artificial reverb with a shorter decay time at low frequencies.
- Record in a concert hall with less low-frequency reverberation.

RUMBLE

Sometimes during acoustic recording sessions, rumbles from air conditioning, trucks, and other sources can be picked up and show in quiet passages. Try the following to clear this up:

- Check the hall for background rumble problems.
- Temporarily shut off the air conditioning.
- Record in a quieter location.
- Use a high-pass filter set around 40–80 Hz.
- Use microphones with limited low-frequency response.
- Mike closer and add artificial reverb.

DISTORTION

If your acoustic recording has distortion, try the following:

- Switch in the pads built into the microphones (if any).
- Increase the mixer input attenuation (turn down the input trim).
- Check connectors for stray wires or bad solder joints.
- Avoid sample-rate conversion.
- Apply dithering when going from 24 to 16 bits.

BAD TONAL BALANCE

Bad tonal balance expresses itself in a sound that is too dull, too bright, or colored. If your recording has this problem, follow these steps:

- Change the microphones. Generally use flat-response microphones with minimal off-axis coloration.
- Follow the 3:1 rule. To prevent phase interference between microphones that are mixed to the same channel, the distance between microphones should be at least 3 times the microphone-to-source distance.
- If a microphone must be placed near a hard, reflective surface, use a boundary microphone on the surface to prevent phase cancellations between direct and reflected sounds.
- Adjust equalization. Try a spectrum analyzer/equalizer such as Harmonic Balancer (www.har-bal.com).
- Place the microphones at a reasonable distance from the ensemble (too close miking sounds shrill).
- Avoid microphone positions that pick up standing waves or room modes. Experiment with small changes in microphone position.

This chapter describes a set of standards for good sound quality in both popular and classical music recordings. These standards are somewhat arbitrary, but engineers and producers need guidelines to judge the effectiveness of the recording. The next time you hear something you don't like in a recording, the lists in this chapter will help you define the problem and find a solution.

FURTHER READING

This edited Chapter is from Bartlet, B. and Bartlett, J. (2005). *Practical Techniques*, Focal Press, Oxford.

Chapter 14
Hints and Tips

Craig Golding and Russ Hepworth-Sawyer

In This Chapter

In this chapter, Craig Golding with Russ Hepworth-Sawyer offer some bits of wisdom before hitting the studio.

MUSIC MATTERS ...

Every producer and engineer can offer a different range of skills, and whether you read music or not, being able to follow or refer to some kind of notation or lead sheet whilst tracking can make a big difference to a recording session. Whilst there might not be a particular necessity for a producer or recording engineer to have some form of notation or chord chart in front of them, this can often have its advantages and can give the artist a new level of confidence in the "musical" abilities of such people.

The ability to communicate clearly and concisely about certain musical phrases, chord changes or song sections may help with the flow of the tracking process and ultimately end in a very productive session. Of course much of this will depend on the musical skills of the artist or band and in some cases this may not be the most suitable aid to communicate musical issues. However, if you ever employ the skills of a session musician then having even a basic ability to communicate in a musical way will put you in a stronger position, and for some projects this may make the difference between getting "the job" or not. It's worth considering. ...

GET BY WITH A LITTLE HELP FROM YOUR FRIENDS ...

Having spent many hours, weeks or even months producing and/or mixing a project it is often difficult to maintain a critical distance from your work. As has been previously mentioned this often manifests itself in the inability to see (or rather hear!) things from an objective perspective. You may have often read or heard that listening to your songs (and your mixes of them) in different environments and through different systems is useful, however, why not try them out on differing sets of ears!

When you get to this stage, being able to hand your "final mixes" over to a trusted (and honest) friend(s) can be extremely helpful and enlightening. For many, this may not be the most natural or comfortable thing to do but it can be extremely beneficial to hear the objective comments they make. Of course, it helps if this person/people have some knowledge of music/music production; however, it is by no means a requirement and non-musicians/engineers can and will offer as valid an opinion and/or comments.

This method of "objective" assistance can be particularly useful when deciding the running order of tracks. Having thought about the emotional architecture of the album and deciding which track should go where, it can be very helpful to hear the comments from a listener who has had no previous association with the project. The song that you have always thought would make a great album opener may not be their first choice, for example. The key here is to find out why this is the case and decide as to whether their points are strong enough for you to change your original decision. You may find it useful to make a list of preference with regard to the running order of your tracks, if your "objective listener" does the same then your respective "lists" can be compared and any discrepancy discussed. If you are in agreement, then the confidence gained at this stage in the project can be very rewarding!

TUNING!

Tuning and intonation of instruments are fundamental issues that really should not require an explanation in terms of their importance when recording. However, it is surprising how many people tend to neglect this area and allow out of tune takes to remain in their recordings. It is true that in today's world of software plug-ins such as Celemony's "Melodyne" and Antares "Autotune" producers/engineers can patch up takes which at one time would have needed to be recorded in order to make it to the final mix. Nonetheless, the use of such tools should be used sparingly if a "natural" and "transparent" sound is to be reproduced. (At this point it is perhaps important to point out that such plug-ins tend to work best on vocal parts although some success may be experienced on other instruments as well.)

It is important not to confuse tuning issues with tonal characteristics. For example, when recording guitar an "organic" sound does not mean that the guitar should be out of tune! There are many ways of achieving a low-fi, clean, underproduced sound without being lazy in the tuning department!

The tuning issue should also be considered in today's climate of "location recording" or rather the ability to record whilst being mobile or on the move. As has been previously discussed in the chapters of this book, the ability to record "demo" ideas that then find themselves in the final mix is becoming increasingly common. With this in mind failing to take a small amount of time to tune properly could mean that an excellent idea/riff from a rehearsal or impromptu recording session cannot be used as the tuning/intonation is out.

Tuning should also be extended to drums. So often these are overlooked and can significantly improve a mix. Ask the drummer to tune the snare and toms according to the track and see how it improves the overall sound of your recording and mixes.

TIMING

On a par with tuning, timing or rather rhythmic accuracy is one of the production basics. The ability to nudge, quantize, humanize, stretch and generally fiddle with timing makes today's Digital Audio Workstation (DAW) a powerful production tool. Many users may opt to use these facilities to "fix" a less than accurate performance. (Beat detective for example!) You may assume that this applies to only the percussive elements of tracks, but many instrumental and vocal takes can and do often have timing issues. The important judgment to make here is whether to "tinker" or "re-take." If you are a proficient user of your chosen DAW then you may always opt to "fix" it, as this will save you precious recording time. However, even if this is the case, sometimes due to the nature of the material or musical phrase it is just not possible and there is no substitute for asking the performer/musician to redo the take. This is especially true if the lack of timing accuracy has affected their delivery of the part in some way. It is also worth bearing in mind that adjusting the timing of parts can often require a fair amount of editing, which in turn introduces the need for crossfades etc., all of which can take time. Be honest, if it only takes 30 seconds to do another take but a minute or two minutes to edit, then the course of action is self-evident.

We could not discuss the issue of timing without mentioning the immortal "click track." Many people seem to be somewhat hesitant to use them and many a musician's choice not to is fueled by their fear of losing the groove, or simply the amount of discipline it takes to play to one! Whether to use a click track or not when tracking can depend on a variety of issues. To make the decision you should ask yourself a number of questions. Are you intending to use any MIDI sequencing or virtual instruments with the track? Does the track change time signatures and/or tempo? Does the track need to be "rock-solid" in terms of tempo? (In most cases the answer should be yes!) If the answer is yes to one or any of these then you will need to use a click. Rushing into a recording and not considering this option first is a big mistake, especially when overdubbing rhythmic parts. How does the performer know when to come in after a break if there is no "click" or count-in?

Redoing take after take in order to get the timing right is not only frustrating but tiring, and can bring an unproductive vibe to any recording session. Certainly playing to a "click track" is not always easy and it is worth asking the musicians if they have prior experience of this. Their response will also give you a good idea as to the amount of success they are likely to achieve when playing to one. If you do find that this is an issue, then using a groove or loop that is tempo matched to the song can make a big difference. This allows the performer to still "feel" the groove whilst staying in time.

However, if the performance is tight, less need for meticulous editing should be required. As has been stated in this book several times, preparation and rehearsal of the material will ensure an accurate take of the material with little need for editing.

LYRICS

When tracking both the song and overdubbing the vocals, it can be hugely invaluable to have copies of the lyrics at hand. Ideally photocopy or print off several copies. Consider using the highlighter technique when tracking vocals:

- Ask the artist to provide you with numerous (preferably typed) lyric sheets.
- Take three colors of highlighter.
- Use the first (lightest color) to denote any areas which have issues with them in the first take.
- Use the second lightest color on the same sheet to denote second pass issues. We hope there'll be less, but not always.
- Use the darkest color to go over with in the third pass.

Using this method, one can quickly see the "hot" spots needing attention. Keeping fast notes can relieve you to concentrate on drawing the best vocal performance possible from the musician. Simply cross out the lines as you comp that perfect take together.

MIX EVALUATION

There will be down points from the recording or mixing sessions, whether that be for a cup of coffee or a longer break. Breaks are excellent to rest your ears and rejuvenate you for a refocus at another time. Upon returning from a break, your ears are at their most objective and it is at this point that a listen back is most useful in determining issues and the next job on the list.

Developing a personalized system by which notes can be taken quickly without drawing too much focus away from listening is recommended. Figure 14.1 is an example but you should use whatever system you develop.

The symbols in this example make the notation quick and easy to take down on a first pass, especially if there is a lot to notate in a short space of time. The symbols here show a "–" for timing issues, "|" for tuning and "O" for any cohesion

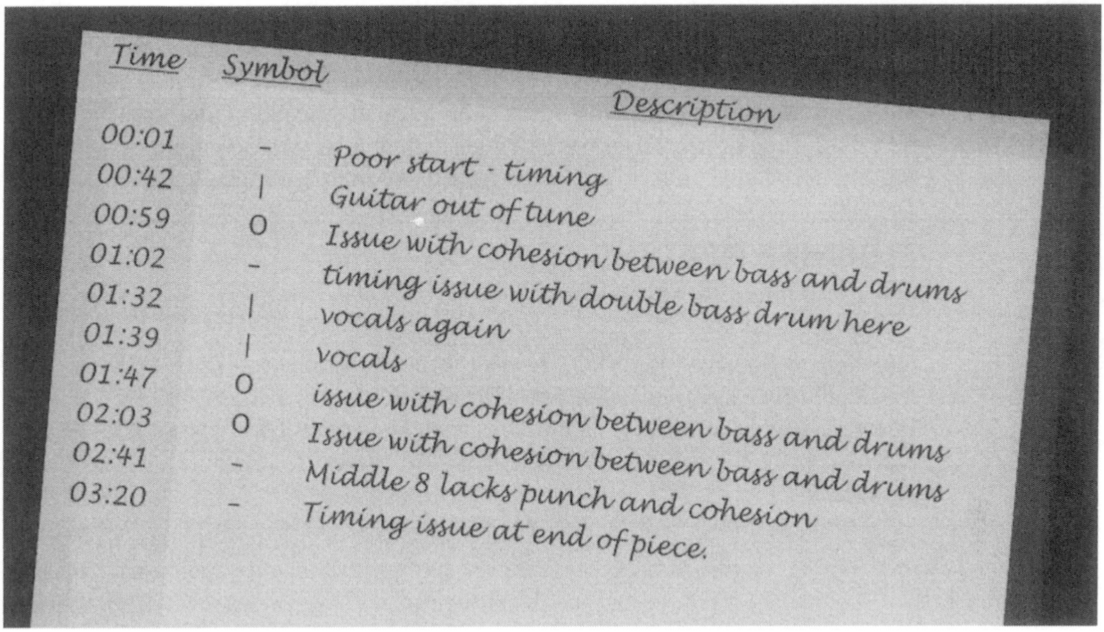

Time	Symbol	Description
00:01	–	Poor start - timing
00:42			Guitar out of tune
00:59	O	Issue with cohesion between bass and drums
01:02	–	timing issue with double bass drum here
01:32			vocals again
01:39			vocals
01:47	O	issue with cohesion between bass and drums
02:03	O	Issue with cohesion between bass and drums
02:41	–	Middle 8 lacks punch and cohesion
03:20	–	Timing issue at end of piece.

FIGURE 14.1
Having a personalized method by which notes about specific issues in a mix can speed up workflow. On the first pass, symbols can be used to note there is an issue, and descriptions can be added later on subsequent passes.

issues needing attention. Latter passes can be used to embellish the descriptions if necessary.

If you are running the DAW, then it will also be possible say with ProTools or Logic to enter markers and simply place the "–" or "|" or "O" in and come back with further descriptions on a second listen.

MASTERING IS FOR MASTERS!

The mix is loud, the vocals have presence, the bass has clout, in fact everything is sounding pretty hot. You then send the project off to the mastering engineer with the instructions to make it louder, tighter, bigger, brighter and hope for that magical polish that mastering engineers add to mixes. You are then disappointed when/if this doesn't happen.

This is not to put down the abilities/job of the mastering engineer, but they can in reality only do so much and have to work with what they are given. If you are thinking of having your project professionally mastered then putting huge amounts of EQ, compression and/or limiting on the mix will make it more difficult for the mastering engineer to achieve the desired outcome. This is a little like asking someone to design the packaging for a product that has already been shrink-wrapped! In fact it is more preferable if you do no "holistic processing" at all. Remember you are paying for a person who has years of experience and knowledge to do this, so let them have room to work and you should be pleased with the results.

You may think that mastering doesn't make a great deal of difference and that it seems like a rather expensive way to add "polish" to a project. However, if you are working on a professional product then there should be no question as to the value of this process and therefore the sense behind leaving the job to the "masters." It can be a hard thing to hand over your lovingly crafted mixes, however, they (the mastering engineers) are "professional" for a reason and in the end it's a matter of trust.

FURTHER READING

Critch, T. (2005). *For Cleaner, Brighter Tracks, Recording Tips for Engineers*. Focal Press, Oxford.

Huber, D. & Runstein, R. (2005). *Modern Recording Techniques*. Focal Press, Oxford.

Izhaki, R. (2008). *Mixing Audio*. Focal Press, Oxford.

Rumsey, F. & McCormick, T. (2006). *Sound and Recording*. Focal Press, Oxford.

Stavrou, M. (2003). *Mixing With Your Mind*. Flux Research, Australia.

Index